Global Environmental Change Science: Education and Training

NATO ASI Series

Advanced Science Institutes Series

A series presenting the results of activities sponsored by the NATO Science Committee, which aims at the dissemination of advanced scientific and technological knowledge, with a view to strengthening links between scientific communities.

The Series is published by an international board of publishers in conjunction with the NATO Scientific Affairs Division

A Life Sciences B Physics	Plenum Publishing Corporation London and New York
C Mathematical and Physical Sciences D Behavioural and Social Sciences E Applied Sciences	Kluwer Academic Publishers Dordrecht, Boston and London
F Computer and Systems Sciences G Ecological Sciences H Cell Biology I Global Environmental Change	Springer-Verlag Berlin Heidelberg New York London Paris Tokyo Hong Kong Barcelona Budapest

PARTNERSHIP SUB-SERIES

1. Disarmament Technologies	Kluwer Academic Publishers
2. Environment	Springer-Verlag
3. High Technology	Kluwer Academic Publishers
4. Science and Technology Policy	Kluwer Academic Publishers
5. Computer Networking	Kluwer Academic Publishers

The Partnership Sub-Series incorporates activities undertaken in collaboration with NATO's Cooperation Partners, the countries of the CIS and Central and Eastern Europe, in Priority Areas of concern to those countries.

NATO-PCO DATABASE

The electronic index to the NATO ASI Series provides full bibliographical references (with keywords and/or abstracts) to about 50000 contributions from international scientists published in all sections of the NATO ASI Series. Access to the NATO-PCO DATABASE compiled by the NATO Publication Coordination Office is possible in two ways:

- via online FILE 128 (NATO-PCO DATABASE) hosted by ESRIN, Via Galileo Galilei, I-00044 Frascati, Italy.
- via CD-ROM "NATO Science & Technology Disk" with user-friendly retrieval software in English, French and German (© WTV GmbH and DATAWARE Technologies Inc. 1992).

The CD-ROM can be ordered through any member of the Board of Publishers or through NATO-PCO, Overijse, Belgium.

Series I: Global Environmental Change, Vol. 29

Global Environmental Change Science: Education and Training

Edited by

David J. Waddington

Department of Chemistry, University of York,
Heslington, York, YO1 5DD, UK

Springer

Published in cooperation with NATO Scientific Affairs Division

Proceedings of the NATO Advanced Research Workshop concerned with education and training in global environmental change science, held at the University of Western Ontario, Canada, June 4 – 8, 1994

ISBN 3-540-59307-1 Springer-Verlag Berlin Heidelberg New York

CIP data applied for

Printed in Germany

Typesetting: Camera ready by authors/editors
Printed on acid-free paper
SPIN 10466216 31/3136 - 5 4 3 2 1 0

This work would not have been realised without the original idea and preliminary work of Max Beran.

PREFACE

A workshop was held in June 1994 at the University of Western Ontario, London, concerned with education and training in global environmental change science. It was principally devoted to post-school education from first year university to postgraduate levels.

I have arranged the papers and the results of discussions in the following order:

Section 1: Introduction

These papers set the scene. It begins with one by Beran whose original idea the workshop was. Schneider outlines various ways in which interdisciplinary courses in general and global change science in particular can be set up. Fyfe looks at the wider issues of global change that affect us all. There are then papers which examine major problems (Leal Filho; Harger and Troost) and programmes to meet them (Carter; Fisher).

Section 2: Global Change Science Courses

These papers are concerned with courses concerned with various aspects of global change courses. They include contributions from nine countries at different levels of secondary and university education.

Section 3: Resources for teaching Global Change Science

New resources are needed for effective courses, particularly those in which students can participate actively. There are wide selections both in terms of topics and teaching strategies, most of which are interactive.

Section 4: Workshop discussions

The workshop participants split themselves amongst three groups, one to discuss early undergraduate courses, another with senior undergraduate courses and the third with postgraduate courses. Their discussions are brought together in this section and common themes are identified.

I had nothing but tremendous help from the participants, my colleagues on the Workshop committee, the NATO Scientific and Environmental Affairs Division office in Brussels, those who helped with all the arrangements at the workshop (particularly Marilyn McMahon) and the embattled secretaries at York (Nancy Newton and Sandra Wilmott). Thus, any mistakes in editing the papers and discussions must lie with me.

York, 1994 D J WADDINGTON

TABLE OF CONTENTS

SECTION 1: INTRODUCTION

Education and Global Environmental Change

Stephen Schneider
Institute for International Studies
Stanford University
Stanford CA 94305, USA
and the National Center for Atmospheric Research
Boulder, CO 80307, USA

Joseph Stoltman
Department of Geography
Western Michigan University
Kalamazoo, MI 49009
USA

David Waddington
Department of Chemistry
University of York
York YO1 5DD
UK

Causes of *Global Environmental Change* fall into two major categories: natural variability and anthropogenic contributions, the human component arising largely as a result of vast numbers of people demanding high standards of living and using technology and resources to support their efforts.

An understanding of the processes leading to global change requires detailed knowledge of the natural and social sciences, and a systems approach. This in turn requires cooperation amongst teachers and researchers across the traditional boundaries in education. These boundaries are between subject areas, and because global environmental change is a multidisciplinary problem, a range of potential solutions and an understanding of the complex issues often spring out of the interface of knowledge among the disciplines.

Further, not only must the traditional boundaries of the subjects be broken down, so too must those boundaries between school, undergraduate, and postgraduate education and training.

It will be seen that these two themes are emphasised many times during this book, in terms of the training of global change scientists, in educating school and university students and in providing the public with the knowledge essential to an understanding of the environment. While earth and physical sciences serve as the information and research base, the nature of the issues in global change bridge nearly every component of the curriculum. Thus, the issues in global change require an educational structure that integrates the natural and social sciences, as well as the humanities and professional studies such as business, law, and medicine.

NATO ASI Series, Vol. I 29
Global Environmental Change in Science: Education and Training
Edited by David Waddington

Why this workshop?

Global change problems are an important and timely focus for research and education and they are indeed being tackled urgently. However, the training brought by most scientists to global change problems consists of first - and often second - degrees in a single scientific discipline, so that the global perspective must be picked up on the job. There are few opportunities for formal training. This problem was recognised during the initial workshop of the NATO Special Programme for the Science in Global Environmental Change (Corell *et al.* 1991), which recognized that the multidisciplinary nature of global change science was a significant obstacle and that NATO had a rôle in assisting with the training and education of the coming generation of global change scientists. The panel in charge of the Special Programme proposed that conversion courses at the masters degree level could be a possible way forward to provide conventionally trained scientists with the necessary global outlook.

A preliminary step prior to holding the workshop was to gather information on existing courses in NATO countries. Enquiries were addressed to research and educational agencies, and centres were visited. From this survey emerged a list of potential participants. However, it was soon apparent that, although there was healthy activity in specific subjects, there were few interdisciplinary systems-oriented scientific research activities or educational programmes. There was, however, an encouraging sign that at least some educational programmes were emerging which attempt to provide students at many levels with a familiarity of the "earth-as-a-system" concept.

These preliminary enquiries also revealed that there was a large measure of interest among educators in having information brought together in a reference volume, which outline global change curricula and resources and experiences of course-providers.

Following detailed correspondence with global science educators over a period of 18 months, it was decided that it was time to bring some people together to share experiences and to chart ways in which global science education at undergraduate and postgraduate levels could be enhanced, and, through this book, to share experiences with others.

The Workshop

Twenty five scientists and science educators from 11 countries agreed to participate in the Workshop. Although the original aim was for the first workshop to identify some of the resources in university postgraduate courses, it was soon apparent that we needed to look at the educational issues below the masters degree level.

During the workshop, it became apparent that in many countries environmental concerns have had little impact on school and first degree science courses. Problems associated with global change education in degrees in single subject disciplines are exacerbated by courses at school and pre-doctoral level, which in many countries are taught in terms of specialisms defined by boundaries set a century ago. This led workshop participants to widen their brief because the training of global change scientists, who have the necessary global perception or requisite combination of knowledge, needs underpinning during first degree courses. With no introduction to the bigger picture of the environment and its relationship to humans, students are not excited about the relevance of the study of science, and we as a society are losing opportunities to further the pursuit of science careers. Getting students interested and committed will in turn contribute to increasing the supply of global change scientists, who have the necessary global perception or requisite combination of knowledge. Further, the lack of a broader science education exacerbates problems associated with a public ill-equipped to face the rapidly changing environment in which we live and is antipathetic to solutions proposed for global change problems. Increasing science literacy through the use of environmental issues will enhance the abilities of the public to understand and participate in environmental policy-making efforts. This broader perspective is reflected throughout the report.

Members of the group had a very wide range of experiences in the education and training of both scientists and non-scientists at university level in various aspects of global change science. They addressed the problems of introducing new and modifying existing courses by having a series of plenary sessions with inputs from participants who outlined the work done in their own institutions and regionally. These are described in Section 2.

The majority of the time was spent in small groups tackling specific issues at three levels of education and training:

(i) senior secondary and freshman university courses

(ii) senior undergraduate university courses

(iii) postgraduate courses

The results of their deliberations are incorporated in Section 3. It was possible to compare experiences, define common objectives for global change educators at many levels, examine the balance between breadth of subjects including various detailed content of specialized topics, identify obstacles to mounting interdisciplinary courses, and, most importantly, fashion a set of recommended solutions to these obstacles.

The differences in outlook of the participants stemming from their different educational systems provided a salutory framework against which we worked, a framework that, all the time, made us think about *flexible* solutions and not solutions that would only apply to one country. There was underlying agreement that there was an urgent need to create an adequately trained talent pool and a global environmentally literate society, which together would provide the foundation for sustainable development.

There was a realisation that representation at a NATO workshop was only from developed countries, and that a vital ingredient was missing: the experience of scientists working in the developing nations. Nevertheless, we had a valuable input from Unesco (Troost *et al.* 1994; Harger and Troost 1994) and a paper from Leal Filho (1994). We did not examine specific ways of educating adults - whether they be policy makers, or members of the general public - either by *formal* or *informal* means, e.g., via the media. We concentrated primarily on the education and training of scientists for research and of those scientists who wish to have a strong element of global change science in their teaching courses at the freshman and senior levels, both in terms of a general education and for development of scientific knowledge and skills.

During the meeting, participants used the same terms with slightly, but still significantly, different meanings. We therefore agreed that we should try to adopt a common terminology so that the reader knew precisely how we were using specific terms. The list of terms was prepared by three participants: Max Beran, Henrietta Lidiard, and Stephen Schneider.

Earth System Science and the Science of Global Environmental Change

The fundamental concept is that of the earth as a dynamic system which has experienced change over all timescales. This is studied as *Earth System Science.* It has been shown that man's activities have the capacity to perturb this global system, so the *Science of Global Environmental Change* is the study of the natural global system focusing on anthropogenic influences and their environmental and societal consequences. This is often shortened to *Global Change Science.*

At present, scientists are focussing on subjects such as climate change, ozone depletion, the loss of biological diversity, degradation of terrestrial and marine environments, and desertification, but the workshop participants were aware that these are but some examples of current and fundamental concern. There are two terms used in the discussion of earth-systems which need defining. These are *linkages* and *interaction.*

Linkage refers to connections among various sub-elements of a system. For example, the ocean provides a heat source for the atmosphere and thus air and sea are linked. *Interaction* implies linkage in both directions among sub-elements of a system. For example, the ocean provides a heat source for the atmosphere, but, for time scales longer than weeks, atmospheric winds and transparency to solar heat feed back on ocean temperatures. Thus, on these scales, there is an air-sea interaction.

Finally, two terms are used to describe the way global change scientists are educated and trained. These are multidisciplinary and interdisciplinary. We distinguish between these in the following way: *Multidisciplinary* refers to knowledge from multiple disciplines derived from disciplinary methods, practices, and paradigms, independent of problems such as global change. *Interdisciplinary* is used to denote the integration of knowledge from multiple disciplines combined in an original synthesis, in order to explain the behaviour of a complex system or to address a problem of practical significance.

References

Correll RW and Anderson PA (eds) 1991 Global Environmental Change. NATO ASI Series Global Environmental Change, Springer Berlin 1 : 264

Harger JRE and Troost D 1994. This book

Leal Filho W 1994. This book

Troost D, Robinson IS and Blackburn D. This book

Evolutionary Organizational Models for Interdisciplinary Research and Teaching of Global Environmental Change

Stephen H Schneider
Institute for International Studies
Stanford University, California and
National Center for Atmospheric Research
Boulder, Colorado

Environmental science, technology, and policy studies require an integrated, interdisciplinary approach that needs to draw together talents from diverse academic disciplines, and from industry, government, and non-government organizations (NGO's) (mostly non-profit advocacy groups), and must offer incentives to participants to produce team-oriented, problem-driven, interdisciplinary research and teaching. The challenge, of course, is how to encourage this and maintain high quality standards, given the traditional disciplinary and/or departmental structures at most universities, national laboratories, and funding agencies. Some individuals have thus suggested that we need to re-examine and modify these traditional structures to ensure they are not impediments to quality, integrated global environmental systems studies.

Linear versus "Organic Systems" Structure for Academic Institutions

Figure 1 is my "cartoon" of the typical academic research and teaching paradigm, a linear progression of "knowledge flow" (represented by the arrows on the figure) from disciplinary and multidisciplinary (but rarely integrated) natural scientific work fed into social sciences for impact assessments which may spawn isolated efforts at policy analysis, sometimes by interdepartmental institutes. Humanities expertise enters sporadically, usually in the form of an interested scholar or two becoming involved with a team of global change natural or social scientists. By "multidisciplinary, I mean specialists from various disciplines applying their respective disciplinary paradigms and methods in parallel on restricted aspects of some broad systems problem. By "interdisciplinary" I refer to individuals or teams of diverse specialists which learn enough - individually or collectively - about a range of relevant disciplinary methods and content to combine such multidisciplinary knowledge into an original synthesis that helps to raise understanding of systems phenomena or to solve a real problem (Schneider, 1977). Interdisciplinary work is often at the intersection of disciplines; multidisciplinary work is usually centered within several traditional disciplines. Government agencies and private foundations largely fund the

NATO ASI Series, Vol. I 29
Global Environmental Change in Science:
Education and Training
Edited by David Waddington

spectrum of such research, but, like other "users" of the research who often have to make policy decisions despite the uncertainties, they largely sit on the sidelines once the grant is given and academic research is undertaken.

Teaching is typically concentrated in disciplinary departments, but a broadening of this discipline-oriented approach has long been undertaken at many institutions. There are many examples described in this volume. This approach will be discussed in depth shortly. Outreach of new knowledge from the academic community to the media, the public, and policy makers is sometimes handled institutionally by extension officers, but, as Figure 1 suggests, it is more often provided by mediaskilled entrepreneurs: an unorganized, ad hoc minority the members of which are selfselected "outreach agents" who interpret and "spin" the academic knowledge-base. Such people are typically disciplinary specialists, and, with a few rare exceptions, offer their interpretations and advice from the narrow perspective of their disciplinary paradigms. What is needed, I believe, is to find an organizational vision that leads to institutions which inherently will immerse such specialists in a systems perspective, an experience that could well substantially alter their interpretations.

Figure 2 is a contrasting organic systems paradigm stressing the integrative or "center-like" aspects of global environmental change problems. This figure illustrates the integration and organizational concepts I wish to discuss here. Rather than a linear, disciplinary organization in which knowledge typically flows from physical to biological to social scientists, and eventually to policy assessment teams, Figure 2 offers a circular, problem-oriented, interconnected structure, or an interactive or organic systems approach. The global environmental change issue is subdivided into three primary subsets: (1) the functioning of the global environmental system, (2) disturbance of the environmental system, and (3) the response of the environmental system to global change disturbances. There is feedback among these subsets and a central entity responsible for integration, which is a group also responsible for feeding back to the three sub-problem groups new scientific problems to work on. An input from many traditional disciplines is necessary for describing the functioning of the global environmental system, and likewise for the other two subsets.

Many gaps in disciplinary knowledge will be uncovered by such a problem oriented approach. Thus, rather than competing with disciplinary research

STANDARD ENVIRONMENTAL STUDIES
(Linear Model - Disciplinary to Multi-disciplinary)

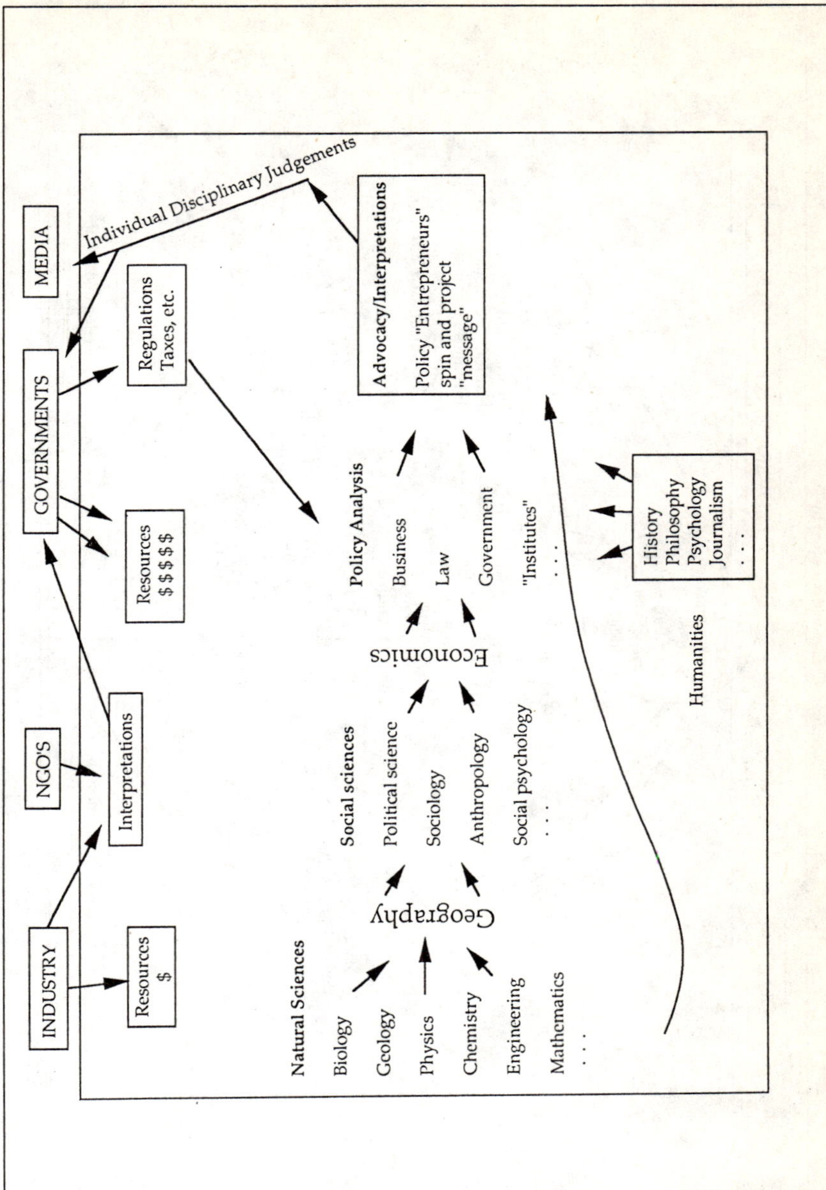

Figure 1

INTEGRATED MODEL OF ENVIRONMENTAL SYSTEMS
(Multi-disciplinary to Inter-disciplinary Research,
Teaching & Outreach Connected by Design)

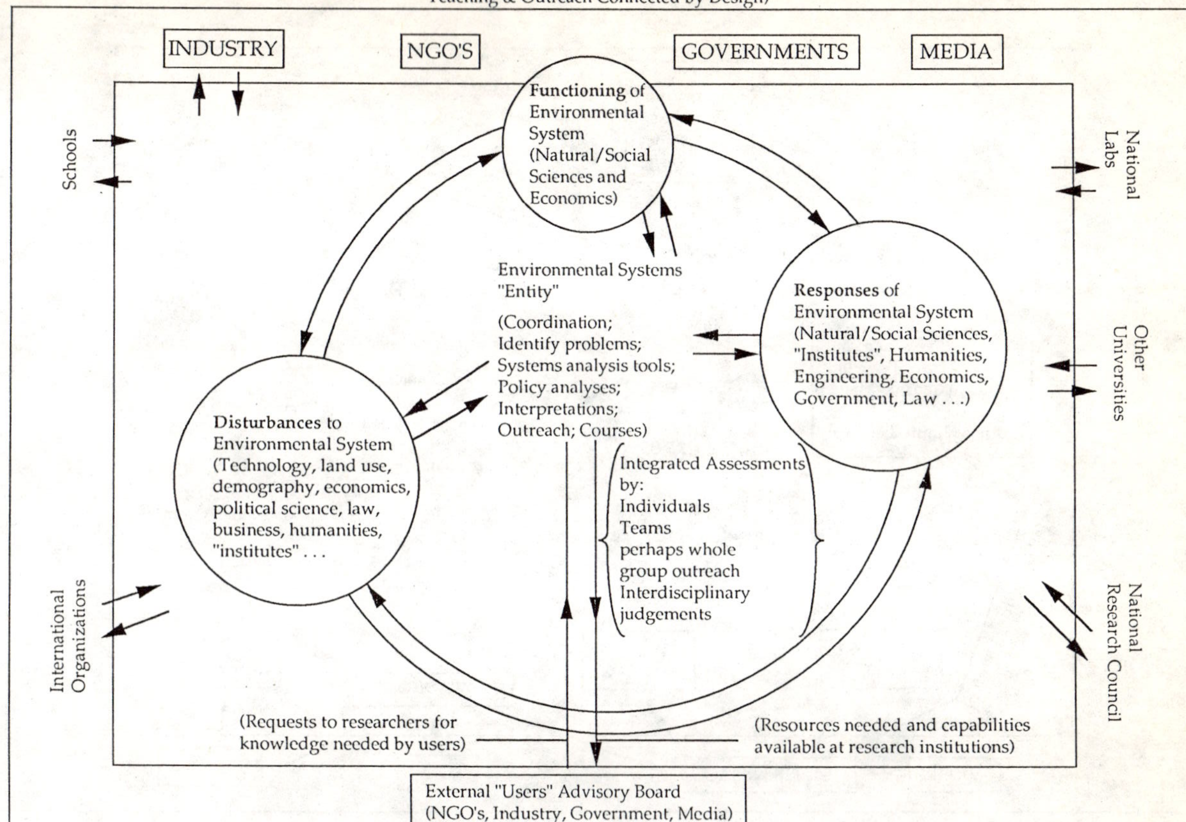

Figure 2

progress, such a systems organization could very well augment it by providing new discipline-oriented tasks for disciplinarians to work on, that are based on overall problem needs and not simply research agendas established in the judgment of disciplinarians as to what is most important for that discipline.

This organic systems structure is designed to unite, evaluate, and reward high quality interdisciplinary research, teaching, and outreach; to coordinate exchanges from multiple disciplines so that an interdisciplinary perspective is facilitated; and to set up formal mechanisms to encourage individuals, teams and (perhaps) whole groups to communicate with the "real world" actors represented on Figure 2 in the "picture frame" margin in the diagram. The three circles reflect a problem-oriented administrative disaggregation of intellectual talent based on environmental systems phenomena, rather than on disciplinary tradition or historic bureaucratic boundaries. Typical disciplinary specialties needed in each sub-problem area are listed in the circles. However, how to organize existing people or groups at an academic or laboratory institution into teams suggested by the circles in Figure 2 is a formidable challenge. In this article discussion will be focused on the intellectual question of whether we would *want* to do this, rather than on the potentially divisive political question of how to effect the transition.

Instead of having practitioners of one discipline pass on their partial view of a systems phenomena to the next discipline, and eventually have the policy implications reported largely ad hoc by disciplinary or special interest-driven policy entrepreneurs (Figure 1), my replacement concept is to evolve a central Environmental Systems Entity (e.g., an integrated department, institute, or some other administrative taxon) to help ensure that actual problems are addressed by having high-quality disciplinary knowledge (available from traditional departments) collected, combined, and focused on problems by the "entity" faculty/staff. Such personnel would include world-leading specialists in the integration process itself. Such scholars may well come from specific disciplines, or be graduates of interdisciplinary programs. Regardless of their origin, they should have demonstrated an ability to seek out specialists with appropriate disciplinary knowledge, and they should help survey and integrate such knowledge in order to enhance understanding of systems phenomena and/or to help solve actual problems.

Evaluating Interdisciplinary Quality

One constraint facing universities considering such a systems approach involves the issue of evaluating interdisciplinary quality. While I could hardly argue with the proposition that researchers should always pursue high quality, it must be recognized that quality evaluation consists of value judgments. Within a university setting, quality generally is defined as any significant quantity of original work that advances the discipline. This definition is perfectly appropriate for disciplinary specialists, but too narrow for evaluation of integration experts. Members of the academic community who define quality under this traditional, narrow definition of disciplinary originality often argue that interdisciplinary programs compromise worth (e.g., Chen, 1981).

I believe that interdisciplinary quality, although substantially different from the disciplinary notion of the concept, is intellectually as viable. Interdisciplinary quality, too, must possess originality. This requirement can be accomplished through an innovative method of combining known materials from various fields rather than within the confines of one field. The materials from various disciplines needed to address a particular interdisciplinary problem must be combined in an original manner that helps to enhance understanding of a systems phenomena or to solve an actual problem. All disciplinary materials used must be up to date and accurate. Furthermore, specialized concepts or jargon must be presented with great clarity to promote cross-disciplinary communication. (These, in fact, have been the criteria for interdisciplinary peer review for the journal *Climatic Change* since its inception [Schneider, 1977]).

Such quality standards can be evaluated, can be enforced, and should be rigorously tested through the peer process - although no individual peer may have the capacity to fully judge the entire nature of the work, nor is that a necessary or a fatal obstacle. The creation of a peer review group for interdisciplinary environmental studies is itself an evolving process, and models of this process will be explored later on in the discussion of Tables 1 and 2. Interdisciplinary work is inherently team-oriented and requires multiple inputs from many quarters. However, over time individuals who become increasingly knowledgeable about the content and paradigms of several disciplines can themselves become interdisciplinary, or in essence, integration experts. These are the kinds of people for whom reward systems are generally not well developed in academe, and for whom funding is often difficult to obtain from traditional sources.

Problem-driven research doesn't exclude disciplinary originality, rather it emphasizes the reverse, requiring more, though problem-focused, disciplinary knowledge. Since that knowledge can be focused on practical problem solving, it should entrain new "customers" for the research products. Thus, this systems model should also provide nontraditional sources of resources to disciplines, that would not likely have been available to curiosity-driven disciplinary researchers from most granting agencies. In short, participation of disciplines in interdisciplinary systems research need not be a threat in the sense of losing out in a zero-sum game, but rather a way of expanding overall support for intellectual activities.

User Interface

In an organic systems model, mechanisms are also built in for interactions with users of research products (for example, environmental activists, industries with environmental connections, elected officials, agency workers, and the public at large) to examine research products and suggest just how these products could - or could not - contribute to improving the knowledge base needed for analysis and ultimate policy choices. In this manner, research relevant to user needs could evolve as an integral component of the center or entity concept. To me, bringing users into research planning is not merely putting "amateurs in charge of disciplines," but rather putting professionals in information use in charge of determining what kinds of information they think they need. Then integration experts at the interface of the user communities and the disciplinary academic and laboratory communities can help to translate what is needed at the policy end into what should be produced at the science end. Quite bluntly, most disciplinary investigators are not qualified by themselves to make that judgment without the input of broader groups, particularly those at the front lines of problem solving. Neither are the problem solvers qualified to dictate the details of what disciplinarians should do, but certainly the former should point out what policy questions are in need of more scientific information, from which integrationists and disciplinarians together could work out focused research agendas that are appropriately multidisciplinary and/or interdisciplinary.

Another vital function that an environmental systems entity could perform would be to establish a place for summer studies, for regional field work, for research fellows to congregate, for media interested in covering global change topics to be put in touch with individuals knowledgeable in various aspects of the issues (rather than the random and/or biased coverage typical of most polarized global change debates), and in which public outreach could be advanced.

Evolution, Not Revolution

In essence, what I am calling for is not the wholesale replacement of the disciplinary paradigm, or its traditions of strict quality review standards for disciplinary research, with something different, but rather the augmentation of standard practices to include problem-driven, interdisciplinary integration as part and parcel of the total academic/laboratory research enterprise. I believe that this evolution toward broader research and teaching paradigms will *not* reduce the quality of disciplinary work. On the contrary, I think that it will expand the opportunities and funding sources, and that evolution of systems organization will not necessarily replace most disciplinary departments, just as biological evolution does not necessarily replace some existing species. Rather, such an evolution will allow analysts and synthesists to address global change problems at the scale they exist (the "problem scale"), a scale not constrained by boundaries and methods that happen to fall into historic entities or disciplines that were created for different purposes.

None of this obviates the need for disciplines. I envision that only a small percentage of overall academic activities would be in the integrationist category, because integration expertise is hard to find, will take time to develop, requires reorganization of training and rewards systems, needs to evolve a peer group, and simply isn't ready to take over a very large fraction of the effort, if quality is to be maintained. Interdisciplinary work cannot function without a high quality base of disciplinary knowledge, but an institution of higher learning cannot efficiently contribute to the solution of pressing environmental problems by retaining a discipline-only, traditional organization and rewards structure. We need both, and there is no time to waste. We simply cannot afford the luxury of waiting for such new structures to emerge bit by bit out of present institutions over the next century, just as specialized disciplines replaced classical training and organizations a century or two ago. We need to be creative and active in accelerating that evolutionary process by perhaps an order of magnitude to address pressing environmental problems whose solutions are already overdue.

Evolution of Global Change Research Organization

Most academic or laboratory research is not organized like the systems model proposed in Fig. 2. However, organizational changes have evolved at academic institutions in an attempt to encourage multidisciplinary and sometimes interdisciplinary activities to take place. Table 1 offers an example of typical organizational issues involved in enhancing global environmental change research activities.

Table 1

Organizational Steps or Issues for Global Environmental Change Research
* Multidisciplinary seminar - to learn who is interested and in what
* Multidisciplinary committees - plan more formal integrative activities and supervise Ph.D. students
* Interdepartmental Institutes (typically not permanent)
- Multidisciplinary faculty with partial appointments in disciplinary departments and institutes
- Senior fellows: non-tenure track but full-time institute appointees
- Development of peer groups capable of evaluating interdisciplinary quality

This NATO volume is dedicated primarily to educational rather than research problems encountered in global change activities. However, research activities typically are given higher priorities than global change educational activities in most of the more prestigious universities, so experience gained in analyzing problems and in finding solutions to interdisciplinary global change research aspects should carry over to dealing with global change educational improvements. Furthermore, the best global change teachers are probably close to evolving research findings.

Returning to the organizational steps in Table 1, the process of using organizational change to encourage global change activities typically begins with the creation of a multidisciplinary seminar. Since the process of establishing networks of multidisciplinary scholars, students, and staff to form multidisciplinary research teams requires a progressive recognition of the problems, practices, and ideas from a variety of disciplines, the formation of multidisciplinary seminars is one of the best ways to build that community of scholars. Talks given by faculty from various departments or from visiting faculty, scientists, decision makers, or people from business and environmental groups from the outside, often serve as a focal point for this activity. Frequently, as interpersonal relationships build around multidisciplinary problems like global change, some subset of faculty, visitors, students, and staff often form the next element on Table 1, multidisciplinary committees. Such groups often plan for more formal integrative activities or can supervise Ph.D. research, evolving to become part of the peer group needed to sustain interdisciplinary quality, as discussed earlier. Such activities could include joint proposal writing, summer studies in multidisciplinary problems, planning for new courses, development of an interdepartmental institute or center, and ultimately, attain a vision for more permanent new departments and/or schools, whose focus would be, in our example, interdisciplinary global environmental change efforts.

Interdepartmental institutes typically are not permanent, hard-money-funded organizational entities at most universities, because they do not have the historical or administrative clout of schools or disciplinary departments. They are typically staffed by multidisciplinary faculty, most of whom may have partial appointments in such institutes, but most of whom have permanence or tenure status in disciplinary departments. Sometimes institutes hire full-time staff, such as senior fellows or research professors, but in many - probably most - current cases such positions are pejorative in the sense that they are not tenure-track faculty slots with full security, privilege, and influence associated with tenure status. In some cases junior faculty (those not yet tenured) are often nervous about participation in such interdepartmental institutes, particularly if their work is interdisciplinary. Such activities can reduce the likelihood of promotion to tenure for junior faculty whose peer evaluation may be made solely on disciplinary standards. Furthermore, they often will be compared to the productivity and disciplinary originality earned by colleagues who may not be splitting their efforts among both disciplinary and interdisciplinary activities.

The tenure issue is explicitly raised next on Table 1. Most tenure-track appointments in academe, or so-called "senior scientist" appointments without term in research laboratories, largely reside in disciplinary departments. Sometimes creative, interim solutions have been attempted across organizational sub-units, such as that at Penn State University, in which the Earth Systems Science program controls the funds for new appointments in global change studies, while the departments retain the legal power to advance or deny a candidate achievement of tenure status. Because of the difficulty in modifying existing legal or traditional appointment practices at academic institutions, such "joint implementation" compromises are often important evolutionary steps of organizational change needed to enhance the tractability and quality of interdisciplinary work.

Another proposed solution to this difficulty was developed in 1969 at Stanford University in its interdepartmental Human Biology program, and is informally known there as "elastic slots." The logic is that research activities and tenure would remain in traditional departments on the grounds that such departments were best qualified to evaluate research quality, but that multi- or interdisciplinary teaching could be carried out in an interdepartmental program or in institute organizations which do not have permanent or tenure granting status. While elastic slots may be a potentially useful interim step, I believe this is not an optimal - or even appropriate - long-term solution, since the highest quality problem solving or interdisciplinary research will require the integration of research, teaching, and outreach in the same organization; and with quality review being applied to both teaching and research, based on how well the candidate performs in the interdisciplinary context, not how well the candidate performs solely in disciplinary research. Rather than seeing departments as best qualified to evaluate research quality, the opposite is more likely: departments are usually not well qualified to judge the interdisciplinary originality or importance of such research activity, and the basic premise behind elastic slots and the assignment of tenure to departments and teaching to programs, is fundamentally flawed with respect to encouraging sustained, high-quality interdisciplinary activities.

The most difficult organizational change for most current institutions would be the creation of new interdisciplinary departments or schools whose prime activity is research, teaching, and outreach on interdisciplinary, systems-oriented problems like global environmental change. To attract the best faculty (especially at early career stages), such an environmental studies entity needs to be backed up with sufficient funding to offer prospective staff the opportunity of permanence on an equal footing with tenure-granting schools and departments. But a new permanent entity is especially hard to create when the overall institution is not growing, let alone when it is undergoing cutbacks. Such new interdisciplinary departments of environmental science and/or management hardly need to add this fiscal obstacle to the many other associated problems of new paradigms, such as evaluating interdisciplinary quality, providing adequate rewards or reprimands based on quality evaluations, and expecting some measure of effective public outreach or technology transfer as part of the job. The suggestion to obtain hard money to support a new interdisciplinary department (which could mean earmarking a fraction of the endowment of a private university for it or becoming a permanent line item in the budget of a large laboratory) is likely to run into a storm of academic opposition from already established organizational sub-units, each of which is trying to hold on to its share of what it perceives as the stable or shrinking pie. Of course, such opposition is usually couched in terms of quality concerns, not competition for scarce resources.

Support for the program, faculty, and staff of a new environmental entity could also be obtained from federal and local grants, as in the case of disciplinary departments and schools, and perhaps could also, particularly in the early phases, get a major boost in support from private foundations. The latter often like to fund new initiatives, particularly for problem-oriented organizations, but do not typically like to be committed to long-term support programs. The creation of new interdisciplinary, permanent organizations would be the ultimate institutional evolutionary step. Most institutions are more realistically described in terms of the activities distributed across the earlier entries on Table 1, but there are some outstanding examples of interdisciplinary environmental studies organization that are already established, (such as the Energy and Resources Group (ERG) at UC Berkeley and the Earth Systems Science Center at Penn State University). At a 1977 symposium on the interdisciplinary process co-organized by the author, the late anthropologist, Margaret Mead, was invited to comment on this question. She characterized three elements she believed were in common to the creation and sustenance of most interdisciplinary ventures: (1) a charismatic leader with sufficient disciplinary credentials to maintain credibility of the enterprise in spite of doubters from disciplines and with sufficient intellectual clout and appeal to procure the next two ingredients; (2) a group of young, dedicated researchers, who are "analogical" rather than "digital" thinkers: or people who at the margins of their intellect like to make connections; and (3) an identified primary funding source for the activities. (Mead's remarks are described in Chen 1981). When pressed by a panelist who was concerned by how difficult these three criteria are to maintain, I recall her replying simply that two out of three of these characteristics needed to be maintained at all times by successful interdisciplinary groups and all three for some of the time, particularly criterion 1 at the outset of the venture. I think her model fits well the ERG and Earth Systems Center examples given above, my former affiliation with the Interdisciplinary Climate Systems Section at the National Center for Atmospheric Research in Boulder, Colorado, and, on a larger scale, the International Institute for Applied Systems Analysis in Laxenberg, Austria.

The Margaret Mead model for interdisciplinary group creation and survival has been closely matched many times and in many types of institutions, but, distressingly, I have also witnessed the withering of such interdisciplinary activities, particularly when the charismatic leader moved on before the unit could become sufficiently established. Thus, such a model implies a burdensome set of hurdles for new interdisciplinary units that most disciplinary groups do not need to address. It is essential for high quality and sustained integrative activities required by such subjects as global environmental change, that support for the activities and staff be institutionalized and put on a par with traditional disciplinary departments. That means having fair access to funding resources, and not being required to maintain indefinitely and continuously either extraordinarily unusual leadership or outside resources as a prerequisite to continued existence. During the transition from initial establishment of interdisciplinary units to more permanent status, it is probably not unreasonable that Mead's model should apply, but eventually a progression through the steps outlined on Table 1 should lead to a more permanent stave .

Successional Model of Organizational Changes to Sustain Global Change

Let us turn now to the question of improving education and training of global environmental students, scientists, professionals, and the public at large. Table 2 summarizes my experience with the teaching of cross-disciplinary subjects at universities and laboratories. The Table suggests a typical progression or successional model for organizational changes in response to the special demands for interdisciplinary teaching. It is a model of a typical successional pattern in which increasing degrees of integrative activities are encouraged and institutionalized. Like a real biological succession, from which this analogy is inspired, sometimes particular examples do not conform with the usual successional sequence (for example, following an intense forest fire ecosystem recovery typically proceeds from grasses to shrubs followed by rapidly colonizing "pioneer" trees and eventually ends up with a final or "climax" state of succession: a mature old growth forest with a stable mix of species). Sometimes standard successional steps are skipped or the order of progression of changes depends on additional factors that range from the scale or intensity of the initial disturbance up to human interventions, such as replanting a certain set of species and eliminating others. The latter could entirely overwhelm the normal successional pattern. Likewise, some educational institutions may have opted for permanent rather than term appointments for an interdepartmental institute's non-faculty staff, contrary to the more usual situation anticipated in

Table 2, in which such permanence for integration experts does not normally occur until the late successional stage when the group becomes a full interdisciplinary department or school. The model in Table 2 is offered simply because this progression seems to capture many typical situations in academe or laboratories, even if individual examples do vary.

Table 2

"Successional" Model of Organizational Steps for Global Environmental Change Teaching

* Multidisciplinary Seminars (get to know the players and rules of their games; typically takes years for individuals or groups to become interdisciplinary)

* Announce Program and Initial curriculum
  - Form multidisciplinary committee to define program goals, participants, and to define core curriculum and electives

* Curriculum Development
  - "Staple-gun" courses (require some half-dozen pre-existing multidisciplinary courses for students to become an "environment major" by displacing similar number of required/elective courses otherwise needed to be a disciplinary major)
  - Tracks (discipline-like course options for upper undergraduates using pre-existing courses)
  - Team-taught, first course (Cameos by major members of staff with interdisciplinary, problem-oriented context emphasized over the disciplinary content
  - New (or modified) core courses (multidisciplinary core knowledge, with content from various disciplines emphasized over integrative, interdisciplinary context)
  - New tracks (add multidisciplinary course options for upper undergraduates)
  - Senior seminar (to reacquaint content-oriented tracked students with integrative interdisciplinary problem context)
  - Co-Terminal Masters allows enough disciplinary content to be equivalent to a B.S. in a major discipline, but with interdisciplinary context of global environmental change problem solving added on)

* Create Environmental Systems Entity with permanence (new department, center or school with tenure-track faculty; research, teaching *and* outreach are combined and their quality evaluated on interdisciplinary systems or problem-oriented criteria)

* PhDs (see Table 1)

* Career change programs
  - A largely multi-disciplinary M.S. program to acquaint already trained disciplinary students and/or practicing professionals with the content of relevant disciplines and the broader context of environmental problem solving
  - Special post-doctoral or mid-career fellowships for senior scientists to retool in an interdisciplinary department, institute, or school.

The successional teaching model in Table 2 is very similar in its basic features to the research steps suggested in Table 1. Both begin with multidisciplinary seminars and progress toward some kind of integrative, systems-oriented environmental science, technology, and policy entity, that could be a center or institute with permanence, a new department, or a school. However, the steps inbetween are somewhat different, with Table 2 chosen to match the typical experience academic institutions have in developing interdisciplinary teaching programs.

Personal and professional relationships often develop at multidisciplinary seminars - seminars that may initially have been organized for building a peer group for research purposes but which also serve to develop relationships necessary for interdisciplinary teaching. At such seminars faculty, students, staff, and other professionals from the region learn enough about each others' interests and abilities to desire further development of a teaching program in global environmental change. The second step often is the formation of a multidisciplinary committee, which tries to define an appropriate curriculum, identify student interest and possible placement of students after completion of courses; and decides who could bring the program both credibility and official status by carrying it through appropriate channels, such as the university management, an academic senate, or governmental officials. Such a committee often writes proposals to help fund or initiate the program and/or tries to attract charismatic leadership either from within the institution or from outside it.

In some cases the multidisciplinary committee might comprise of faculty not only from disparate departments, but from different schools; whereas in other cases faculty might come from different areas of the same department, but be assembled to fashion a new, broad program within that department. Regardless of which model applies, a typical product of such committees is to define a curriculum for a departmental or interdepartmental program related to global environmental change. As the successional model of Table 2 suggests, there are some common stages in the development of such curricula. In the earliest phases after the program is announced, requirements for a core curriculum, required of all students, and/or electives open to upper-class students, are often listed. The courses involved are often already in the institution course catalog and lead to what I have referred to in Table 2 as "staple gun courses", whereby the multidisciplinary program simply "staples" together existing courses in unique combinations not typically open to disciplinary majors. For example, students enrolling in the new environment program either as a major or a minor (depending upon institutional options) might substitute for disciplinary major requirements, several courses from a variety of departments that involve relevant multidisciplinary global change questions. These substitutions would thus displace a similar number of core courses or electives that otherwise might have been required in order to receive a degree in a specific discipline. The obvious advantage of such an approach is that it allows the student and his/her advisor flexibility in putting together an individualized curriculum which more closely matches his or her perception of what constitutes the knowledge base needed to deal with particular aspects of global environmental change, than might be available in a traditional discipline. The disadvantage is that the "stapled" courses themselves are often aimed at discipline majors, and may provide more detailed or peripheral material than is really necessary for the interests of the global environmental change student. For example, an atmospheric science student wanting to learn enough about leaf physiology to understand how to incorporate biophysical processes into a climate model might wish to have some detailed information on leaf physiology, photosynthesis, and perhaps only a few disciplinary lectures would be sufficient for this interdisciplinary purpose. Yet, the staple gun approach would require the student to take an entire semester or more of a botany class to gain the information legitimately. Likewise, a student trying to understand the development process in China and wishing to compare it to the Victorian Industrial Revolution in Europe or North America might register for a history, economics, or political science course that deals with Western industrial development in the late 19th century. Once again, it is possible that the knowledge needed by that student would be sufficient with half a dozen lectures summarizing the highlights of the issues. It may not be efficient to have the student learn in full semester's depth all the labor relations, social and political conditions, innovations in industrial technology, and other details that appropriately belong in the disciplinary courses dealing with that time of Western industrial development.

In view of the demands for multidisciplinary training and the limitations of a four-year bachelor's degree, it may not be practical for a program to strive for full depth in several relevant global change disciplines, and also adequate breadth by exposing the student to a vast array of unmodified physical, biological, and social scientific courses; as well as policy, history, and values studies - all in a standard bachelor's or master's candidacy! This suggests that in the best teaching programs attempts will be made to try to create new core courses and eventually new electives, the latter to be specifically designed to meet the needs of global environmental change students. This could be done by selecting representative samples of material from existing disciplinary courses and combining them, perhaps in an original way, in specially focused global change classes. Of course, such new courses will require staff to teach them, but sometimes difficulty is encountered because the breadth of expertise to teach these "environment" courses is not resident in the department or even in the program. With luck, such expertise is available somewhere across the university, but then getting people to participate can be a problem, especially if there are few rewards for participation, and frequently multiple instructors are needed. This has the advantage of comprehensiveness; the disadvantage of choppiness. One solution is a single course coordinator whose job it is to smooth the edges between jumps, as lecturers dealing with different disciplines follow each other in sequence.

A first course in global environmental change is an example of this issue. First of all, such a course could satisfy the science requirement for non-science majors in a given university. Recent experience at a number of institutions suggests that vast numbers of students are interested in a comprehensive global environmental change course for their general science education requirement. Such a course could also be a springboard for someone in the program planning to take further courses. Cameo appearances by a multidisciplinary set of top faculty typically dominate such a course, but alternatively, visiting scientists, policy leaders, or industrial representatives can be guest lecturers, and together with program staff provide the breadth needed at the introductory level. Such a course stresses the context of global environmental change and its broad multidisciplinary nature over the content that would be more typical in first disciplinary science classes.

I strongly believe that at the outset of learning about global environmental change, students need a sense of how physical, biological, and social sciences all contribute to understanding global change problems. Beyond that, students need to be aware of the policy making process and understand that human values lead to policy choice. Environment programs that have just begun may have no more than one such new course, but more developed programs often attempt to fashion new (or modify existing) courses as part of the environment core. Occasionally, new electives in the global change program may be offered.

The Stanford University Example

Earth Systems Undergraduates. My own experience at Stanford University provides an example. The Earth Systems (ES) program at Stanford was initiated by biologist Jonathan Roughgarden and geologist Gary Ernst. ES is an inter-faculty, inter-school program managed by a committee of some 18 professors from all across the campus. A beginning course, affectionately called the "grabber," is offered as a core requirement for earth systems majors as well as being offered as an option for fulfilling science requirements for all students at Stanford. Its first year saw an enrollment of about 20 students, second year 40, and third year over 100. In 1994, in less than three years, there were nearly 90 majors in this program, attesting to its strong popularity with the students. The Stanford program leaders believe that physical, biological, and social sciences all must be included in a core curriculum taken by all earth systems majors. This core goes beyond the grabber course taught by "cameos" given by half a dozen highly visible professors and a few outside visitors. Three multidisciplinary courses which stress content over context in three physical, biological, and social scientific areas are required. One deals largely with geology (the "geosphere" course), one with biology (the "biosphere" course), and one with economics (the so-called "anthrosphere" course). These courses are cross-listed between Earth Systems Science and the relevant disciplinary departments, and the syllabus has been modified by compromise arrangements with the professors, his/her department, and the earth systems leadership to try to meet the needs both of disciplinary students taking the course and not involved in the earth systems program (who typically request more depth), and earth systems science majors who want to cover more topics in that sub-discipline than might be typical of the major. Stanford has also required calculus, physics, chemistry, economics, and biology background courses from the catalogue (i.e., the "staple gun" approach). The ES requires the students to take those courses (e.g., physics, economics, and biology) that have more mathematical prerequisites, thereby stressing methodological rigor, regardless of whether the subject is economics or ecology. At Stanford, five "tracks" are available to students (Appendix A). These are discipline-like programs for upper undergraduates, that provide depth and content equivalent to roughly two-thirds of the content that might be obtained by a disciplinary major who had not had the benefit of the added breadth associated with the earth systems program. Currently, the tracks at Stanford are "staple gun-like" in that few new courses have been specifically developed for Earth Systems beyond the grabber and the three crosslisted core courses - although a few special topics classes have been offered on land use problems and the management of energy efficiency. However, should the ES program gain permanence and therefore have the capacity to hire tenure-track faculty directly into the unit, it would be desirable to create new courses and even new tracks. (An energy track would be a high priority in the opinion of this author and member of the ES Steering Committee.) However, such events are later stages on the successional model of Table 2, at least so far at Stanford.

Another element to the Stanford program common to a number of other institutions is a senior seminar. Here, the attempt is to reacquaint students with the interdisciplinary context they learned in their first course but moved away from in discipline-like tracks. Such seminars involve outside speakers, faculty members, or, preferably, detailed presentations by the tracked students themselves that are subjected to the comments, criticisms, and perspectives of their multidisciplinary student colleagues.

Senior Honours. Another approach to dealing with the senior level interdisciplinary integration need is a "senior honors" program in environmental science, technology, and policy. Based on recent Stanford experience (Appendix B), this has been run through the Institute for International Studies (IIS), which is an institute consisting largely of a multi-disciplinary faculty with half-time, tenured appointments in disciplinary departments and half-time, non-tenured Senior Fellow appointments in the Institute. Institute Fellows, as well as faculty with full-time appointments in departments, make up an interschool faculty committee to oversee the program. (Such a committee helps to advance the process of peer group formation mentioned earlier.) A few full-time institute staff with non-tenure "Fellow" status are also involved. External funding for the honors program was obtained for several years of operation. A competitive application procedure has produced a remarkably diverse and high-quality set of students for the honors program (Appendix B). After the first year's experience, the four faculty involved (with only a dozen or so students each year) recognized that even after three years of multidisciplinary training, Stanford students still needed more hands-on experience with the integration process, particularly for the policy aspects of environmental issues. Thus, the first quarter of the IIS honors program is now devoted to the analysis of practical case studies of environmental science, technology, and policy problems (e.g., acid rain, global warming, etc.). Lectures on risk analysis, cost benefit methods, the regulatory context, and communicating complicated issues into the political process through the media were all included. Students are expected to know the difference between "facts" and "values," what testimony is given, and to be familiar with the nature of the decision making process in industrial, federal agency, and legislative bodies.

The next two quarters are devoted to student presentations on the ideas behind their theses, preparation of interim drafts, and ultimately their final defense before faculty and student colleagues. Simply put, as the year goes on, questions and comments from the faculty at honors sessions are deliberately restrained and the students themselves, with diverse background (Appendix B) begin to take over the job of melding themselves into an interdisciplinary unit capable of penetrating analyses of their colleagues' work.

As one of the four faculty in this experimental program, I can attest to the high degree of satisfaction personally derived from association with these hard working and dedicated students, and the amazingly high quality of their work. (Several of their projects already are being considered for formal publication.) On the other hand, in all candor, it is extremely expensive to have student-to-senior faculty ratios like the one in this program (12 to 4), so this model is simply not feasible for the vast majority of students at most educational institutions. That would have been true at Stanford, too, had not a charismatic leader here secured a generous, but temporary, externally funded gift. Nevertheless, what can be readily generalized from this experience is the important need to bring "tracked" students back together on real world problem-solving exercises, and for them to have multidisciplinary contact with their colleagues that will hopefully lead to some interdisciplinary rapport and understanding.

Co-terminal Masters in Earth Systems. One potential drawback of the successional model in Table 2 or of the Stanford examples presented so far is that by the end of the bachelor's program, an earth systems graduate would typically have had less in-depth training than a disciplinary major graduate, although the environmental change student's training would provide a much broader context in which to place environmental problems. In order to increase the competitiveness of such environment program graduates with disciplinary graduates in job markets, Stanford recently created a "CoTerminal master's degree." This option allows a student to receive both bachelor's and master's degrees in earth eystems in five years. The master's has rigorous requirements (Appendix C) for basic, multidisciplinary methodological training in physical, biological, and social sciences, and math. However, a student is then free to pursue further course work in his or her specialized track. Thus, upon graduating in five years, he or she could have depth in disciplinary content roughly equivalent to a B.S. in a discipline, but the ES student likely would have much greater understanding of the interdisciplinary context of global change problem solving. Preliminary experience suggests that this not only puts the "environment" students on equal footing for disciplinary jobs with bachelor's degree disciplinary majors, but gives them a distinct advantage if they choose to work in a variety of environmental fields.

A Permanent Environmental Unit: The Environmental Educational Climax

On Figure 2 an "entity" was proposed in the center of the diagram to house on a permanent basis the integration experts who serve the purpose of being an interface among scientists from different disciplines, and the outside "user community" which needs emerging global change information for practical decision making. Such an entity could be a new department and/or school, or perhaps even an existing institute or center. The factor that puts it at the "climax end" of the successional model on Table 2 is that it would have permanence, with the capability of granting tenure to deserving and quality reviewed staff (with quality review standards based on interdisciplinary rather than disciplinary criteria, as outlined earlier), and would be an institutional home where research, teaching, and outreach can all be practiced and evaluated as a whole. With regard to outreach, it is my belief that scientists involved in global environmental change studies have the obligation to inform their colleagues and society at large in clear and understandable terms about the emerging state of the art in this field. While not all scientists are equally skilled at public presentation or technology transfer, nor do all desire to be involved in the policy process, at a minimum I believe that students trained in global change science should be given some instruction in techniques for clear communication, the nature of media coverage, the use of metaphor, and the need for public information in this area. Then, they, as well as faculty residing in some global environmental change organizational entity, can choose to what degree they wish to emphasize public outreach in their career; but at least they will be informed of the context and have some skills in this area. At a minimum, I would hope that all global change professionals and students would be receptive to the different outreach styles of their colleagues, including those who invest great efforts at this - provided those outreach activities are of high quality and are responsible representations of often confusing and contradictory results.

At this stage, I might add a personal prejudice: that the best teachers of global environmental change will probably be those who are either themselves researchers immersed in interdisciplinary problems, or who are intimately familiar with research and researchers who pursue this kind of work. For this reason I believe that new organizational entities, which are dedicated to integrative activities that unite research, teaching, and outreach, will have the highest quality global environmental change activities.

Post Graduate Training

Finally, two additional categories appear at the end of Table 2 and deserve brief mention. One is Ph.D. education and the second is career change programs for existing professionals or students who are already trained. With regard to Ph.D.'s in global environmental change, there is legitimate debate among those who believe that global change Ph.D.'s should be problem solvers pursuing interdisciplinary research according to the three *interdisciplinary* quality criteria mentioned earlier; and others who feel it is still in the best interests of students to pursue a more narrowly focused, discipline-like thesis, and then to become increasingly broad in the next phases of career growth 'or perhaps to defer such breadth until after tenure. While there are pro's and con's to both positions, my belief is that institutions should not rigidly adhere to either model, but should make flexible decisions based upon faculty and student interests and abilities, and individuals' willingness to take risks. However, in all cases a Ph.D. candidate in global environmental change needs to recognize that there may be suspicions about his/her capacity to do in-depth work. Thus, regardless of whether the originality of the thesis is in the disciplinary component, the interdisciplinary component, or both, a Ph.D. candidate would be well advised to be particularly careful to produce a substantive, useful, and original piece of work.

As for the retraining of working professionals from industry, government, or individual disciplines, who wish to pursue a career shift into global environmental change areas, a number of institutions have already created master's degree programs for such students. What seems essential in these programs is to have a core set of courses that insure that the student has adequate depth in physical, biological, and social sciences to understand the global change context, while at the same time having the opportunity to pursue special topics in more depth. With regard to the Ph.D. program again, there are many examples in many institutions of individual faculty members who work across departmental or institutional lines doing multi- or occasionally interdisciplinary environmental research. Such faculty also are often willing to fund and train a few graduate students on such projects. However, as laudable as this activity may be, I must ask: How many of these still unusual interdisciplinary Ph.D.'s would be hired as entry level faculty members by a university, whose departments are structured by discipline, given that the disciplinary contributions of these young scholars may not be as weighty as other recent Ph.D.'s who stayed fulltime within the traditional disciplinary boundaries and/or published single-authored, basic research papers rather than team-authored, problem-driven publications?

This raises for me an ethical issue: Should we in academics be training such interdisciplinary doctorates if we wouldn't hire them as colleagues? Furthermore, can we do a truly fine job of integrated interdisciplinary teaching when so few of us have become adequately knowledgeable in three or four environmentally-connected, but still distinct disciplines - disciplines that are needed to do interdisciplinary research that is quality evaluated on interdisciplinary rather than disciplinary criteria?

Advocacy Conclusion

It is my belief that only through the ultimate creation of new, permanent, interdisciplinary environmental organizations within academic institutions can such high quality programs be pursued on more than a piecemeal or temporary basis. It seems unimaginable to this "global changeologist" that universities, or other research and educational institutions claiming a significant share of public resources, could willingly turn down the opportunity to examine both intellectually interesting systems problems and socially important questions, such as understanding, predicting, and managing global environmental change, simply because it demands modification to existing institutional structure or because peer prejudice is in the way. Solutions to the formidable obstacles outlined in this chapter and elsewhere in this volume are given throughout the book. Hopefully, models such as those presented here and in other chapters can help managers, scientists, legislators, and educators to develop or expand research, teaching, and outreach efforts in global environmental change. I do not believe that these models necessarily fit all situations, nor should they be rigidly followed. The essence of the development of global environmental studies entities is flexible institutional change fashioned to meet the specific needs of a wide variety of actors and institutions. But the one thing of which I am certain is that current institutions are not remotely adequate to the task of producing sufficient numbers and quality of students, scientists, managers, and environmentally literate citizens to match the urgent needs of a world undergoing rapid and potentially serious global environmental changes.

Acknowledgments

I appreciate the comments and suggestions of Rosamond Naylor and Donald Kennedy on an early draft of this chapter.

References

Chen Robert S (1981) Interdisciplinary Research and Integration: The case of CO_2 and Climate. Climatic Change Volume 3 No. 4: 429 - 447

Schneider Stephen H (1977) Editorial for the First Issue of Climatic Change and General Guidelines for Reviewers. Climatic Change, Volume 1 No. 1: 3 4

Schneider Stephen H (1992) The Role of the University in Interdisciplinary Global Change Research: Structural Constraints and the Potential for Change, an editorial. Climatic Change Volume 20 No. 1: vii - x

Appendix A

Excerpts from Stanford University's Earth Systems Program Materials 1994.

Faculty

The faculty who teach in the program consist of the most outstanding scientists, scholars, and teachers in the Stanford community. The program is a Stanford contribution to the international consensus that environmental sciences and economics must immediately work together. Leaders from much of the world in politics, industry, and academia, together with a huge number in the general public, have called for reconciling environmental values with economic prosperity while preserving individual freedom. The Earth Systems Program at Stanford is one of the few places in the world where students can study the components of both the natural and social sciences in a coherent, contemporary, and rigorous curriculum of environmental knowledge.

Undergraduate Curriculum

Cognate Courses Required for all Majors:

- Biology (choose one):
  - Biochemistry, genetics, and molecular biology
  - Plant and population biology
  - Principles of ecology
- Chemistry
  - Chemical principles
  - Organic chemistry
- Geological and environmental sciences
  - Planet earth or earth history
- Mathematics
  - Calculus and analytic geometry
- Statistics (choose one)
  - Biostatistics
  - Introduction to statistics methods for earth and environmental scientists
  - Theory of probability
  - Statistics for social scientists
- Physics
  - Mechanics
  - Light and heat
- Economics
  - Elementary Economics
  - Economics Analysis
- Computer Programming
  - Programming Methodology

Courses Required for all Majors:

- Introduction to earth systems or introduction to environmental science and technology
- Geosphere
- Biosphere
- Anthrosphere
- Internship
- Senior seminar

Geosphere Track

- Earth minerals
- The oceans
- Chemistry of the earth
- Environmental and geotechnical geophysics, or general geophysics
- Structural and engineering geology
- Three electives, for example; management of geological hazards; the earth's nonrewable resources; volcanology; natural hazards of man; sedimentary geology and petrography depositional systems; structural and engineering geology ; energy resources; groundwater pollution and: environmental problems in the petroleum industry

Biosphere Track

- Biochemistry, genetics and molecular biology
- Cell and development biology
- Plant and population biology
- Principles of ecology
- Evolutionary genetics
- Plant adaptations, or behavioral ecology
- Three electives, for example: plant adaptation; ecosystems of California; biomechanics of intertidal organisms; biochemical evolution; theoretical population genetics; theoretical ecology seminar, or geophysics

Anthrosphere Track

- Economic analysis
- Economics and public policy
- Economics of development, or Third World population perspectives
- Three electives, for example: world food economy; economic development in Africa; analysis of development projects; economics of legal rules and institutions; International economics; U.S. Agriculture and the Environment; antitrust and regulation; economics of natural resources; environmental issues in international relations

Land Systems Management Track

- Water resources
- Environmental earth sciences I
- Environmental earth sciences II
- Environmental earth sciences III
- Three electives, for example: environmental science and technology, or environmental planning; hydrogeology; introduction to Urban Studies; environmental and geotechnical geophysics

Environmental Technology Track

- Water resources
- Mechanics and fluids
- Introduction to Chemical Engineering
- Two of the following:
  - Environmental science and technology
  - Environmental planning
  - Air quality management
- Three electives, for example: open channel and pipe flow; small scale energy systems; water quality control processes; building energy lab; energy resource; technology and economic change or development and technology in Third World; engineering thermosciences; environmental and geotechnical geophysics

Undergraduate Projects

The program offers students the opportunity to augment their formal coursework with supervised field, laboratory, private sector, or advocacy projects of their choice. In addition to providing experience outside the classroom, the project gives students an opportunity to try out potential career paths prior to making the commitment inherent in accepting a career position or entering into a particular graduate program. The project may be: 1) an internship of the student's choice; 2) directed research under the supervision of a faculty member; or 3) participation in one of several Stanford offcampus programs dealing with earth systems themes (e.g., Stanford in Washington, spring quarter at Hopkins Marine Laboratory, etc.). The project is required of all declared majors in the program, and is normally undertaken between the junior and senior years.

The experience should be a challenging and focused project pertaining to problems in earth systems in the area of the student's declared track, involving a time commitment equivalent to 9 units of academic credit.

Each student is expected to submit a written report on the earth systems project to the academic coordinator. This paper should include the objective of the project, the student's role and function while engaged in the project, and the relationship of the project to the student's track within the Earth Systems Program. This could be followed by a techniques/methods section and a presentation of any data acquired. Finally, the student should include a section addressing the implications of the work and possible directions for future work. The report should be 10-15 pages in length. In many cases, reports or documents prepared as part of an internship are themselves suitable earth systems project reports.

Senior Seminar

Each student is expected to participate in the Senior Seminar. As part of the seminar, the student will make a presentation based on the earth systems project report, and lead a discussion of the implications of the work. The presentation should include a historical perspective, the goals of the project, the student's responsibilities, and the results of the work. The student should be prepared to discuss the work with the rest of the class and with students from different tracks, and should be able to place their work within the larger earth systems focus.

Appendix B

Excerpts from Materials for the Goldman Interschool Honors Program in Environmental Science, Technology, and Policy at the Institute for International Studies, Stanford University

Environmental studies at Stanford have expanded rapidly in both breadth and depth during the past 5 years in response to widespread faculty interest and increasing student demand for further research and teaching in the field. Almost 50 faculty members and permanent research staff from the Schools of Humanities and Sciences, Earth Sciences, Engineering, Law, Business, and Medicine are pursuing research directly related to environmental science, policy, and technology. Among them, at least 25 are involved in curriculum development and programs that integrate teaching and research. These efforts are helping to serve three important functions: to provide environmental literacy to all undergraduates; to offer specialized study for some undergraduates; and to develop distinguished research and graduate training programs in environmental science, technology, and policy that span several disciplines.

The Institute for International Studies (IIS) is taking a leadership role in promoting both the interschool and international policy dimensions of Stanford's environmental studies program. The importance of these efforts is underscored by the facts that: 1) environmental problems, such as ozone depletion, loss of biodiversity, water contamination, and acid rain, are not restricted by political boundaries; and 2) the solution to these problems requires he combined efforts of scientists, engineers, economists, and policy specialists. No single discipline is capable of addressing the full scope of any major environmental problem.

The IIS's strategy is to facilitate interschool cooperation, enhance interdisciplinary research opportunities, integrate research and teaching activities, and develop research and internship programs for students in the environmental field both in the U.S. and abroad. During the past 5 years, the IIS's efforts have been very successful in improving communication among faculty in different disciplines and reducing the departmental barriers that often prevent effective interdisciplinary work.

In September 1993, the IIS initiated the Goldman Interschool Honors Program in Environmental Science, Technology, and Policy for undergraduates throughout the university. The program serves as a focal point for uniting widespread student and faculty interest in the environmental field, and contributes substantively to the development of an integrated academic program among departments with courses directly related to the global environment. The goal of the honors program is to facilitate communication among students in different departments and to prepare them for the type of interdisciplinary interactions that they are likely to encounter as professionals following graduation.

The program currently draws four to five exceptional students from each of the three undergraduate schools (Earth Sciences, Humanities and Sciences, and Engineering) into a small group seminar. The seminar meets throughout the year; the first phase consists of faculty lectures and problem sets, and the second phase consists of student presentations of their work. The seminar is led by Lawrence Goulder, Donald Kennedy, Rosamond Naylor, and Stephen Schneider, all of whom have appointments at the IIS. An interschool faculty committee, consisting of 10 senior faculty members from Earth Sciences, Engineering, and Humanities and Sciences, selects the students each year and certifies their theses.

The focus of the program is the completion of a thesis by each student that reflects innovative and individually motivated research. Ideally, each thesis demonstrates intellectual depth in the student's primary field (e.g., economics, biology, earth sciences, or engineering) and some breadth to place the thesis in a scientific, technological, and policy-related context. The theses are oriented around scientific, technological, and policy-related aspects of leading environmental issues of international scope and relevance, such as biodiversity loss, greenhouse gas emissions, toxic wastes, deforestation, degradation of productive agricultural lands, transboundary water pollution, and population growth.

Appendix C

Further excerpts from the Earth Systems Program Materials at Stanford University pertaining to the coterminal bachelor's and master's degrees

The Stanford coterminal degree plan enables an undergraduate to embark on an integrated program of study leading to the master's degree before requirements for the bachelor's degree have been completed. An undergraduate majoring in earth systems may apply to work simultaneously toward bachelor's and master's degrees. The master's degree in earth systems provides the student with enhanced tools to evaluate the primary literature of the discipline most closely associated with the student's track, and allows an increased specialization through additional course work that may include nine credits of thesis research. Finally, integration of earth systems concepts is furthered by participation in the spring seminar, where the master's student is expected to play a leadership role.

Students should apply through the Earth Systems office. They are required to complete an application which includes a statement of purpose; a Stanford transcript; two letters of recommendation, one of which should be from a faculty member of the program; and a list of courses in which they intend to enroll to fulfill degree requirements. Students may be admitted as early as their eighth quarter at Stanford, or after earning 105 units, but no later than their eleventh quarter. Students may either 1) complete 180 units required for the B.S. degree, or 2) complete a total of 15 quarters during which the requirements of the degrees are fulfilled concurrently.

The student has the option of receiving the B.S. degree upon completion of that degree's requirements, or of receiving two degrees concurrently at the end of the master's program.

Four levels of requirements must be fulfilled to receive a M.S. degree:

1) all requirements of the B.S. degree
2) an enhanced set of cognate courses or their equivalent
3) further course work (and/or thesis research), all of which should be at the 100-level or above, and including 18 units at the 200-level or above, leading to further focus within the student's track
4) participation in the spring seminar

The program must consist of a minimum of 36 units of course work and/or thesis research, at least 18 of which must be at the 200-level or above. A more detailed description of the coterminal master's degree may be obtained from the Earth Systems' office.

1. Cognates

The following subject areas should be mastered, if this has not already been accomplished at the B.S. level. Suggested courses are indicated, but others may be substituted with the advisor's approval.

Courses in differential equations, linear algebra, statistics, optimization theory, ecological systems, thermodynamics, geochemical cycling (given by the department of Mathematics, Statistics, Economics, Biology, Chemistry and Earth Sciences).

2. Specialization Within Track

The student must devise a program of study that shows a level of specialization appropriate to the master's level, as determined in consultation with the advisor. At least 18 units must be at the 200 level or above. The program should demonstrate further specialization and focus within the student's undergraduate track.

The program should be developed prior to application to the master's program, and will be part of the application to the Earth Systems Committee. Nine units may be in the form of research, culminating in the preparation of a master's thesis. A thesis is not required for the degree.

3. Integration

Master's students must take part in the spring seminar (ESys 290), and will have additional responsibilities appropriate to the master's level (thesis presentation, modeling problems, etc.), 2 units.

GEOSPHERE TRACK

Required courses:

Chemistry of the earth (4 units); the oceans: an Introduction to the marine environment (3 units); sedimentary geology and petrography (4 units); hydrogeology (5 units) or hydrology (3 units).

Other recommended courses:

Surface waters; low temperature aqueous geochemistry (3 units); sedimentary basins (3 units); sedimentary petrography (2 units); introduction to micropaleontology (5 units); energy resources (3 units); management of Earth Resources (3 units); role of fluids in geologic processes (3 units); introduction to groundwater solute transport (3 units); environmental aspects of petroleum engineering (3 units); plate tectonics (3 units); tectonophysics (3 units).

BIOSPHERE TRACK

Recommended courses:

Genetics (3 units); general botany (3 units); plant adaptations (4 units); ecosystems of California (4 units); systemics and ecology of vascular plants (5 units); plant genetics (3 units); biomechanics of intertidal organisms sum (6 units); plant physiology (4 units); conservation biology (4 units); biology of insects (4 units); biology of birds (3 units); biochemical evolution (3 units); advanced plant systemics (2 units); mathematical models in population biology (3 units); behavioral ecology; theoretical population genetics (3 units).

ATMOSPHERE TRACK

Required courses:

Price allocation theory (5 units) or economics analysis (3 units); economics of the environment (5 units); economics of natural resources (4 units).

Recommended courses:

Population perspectives in the third world (5 units); industrialization, growth, and economic development (5 units); international trade theory (5 units); public finance and taxation (5 units); introduction to political economy (5 units); essentials of political theory (5 units); world politics and the global economy (5 units); advanced ecological anthropology (5 units); public policy design (3 units); environmental law and processes; environmental strategy.

Education and Training of Global Environmental Change Scientists

Max Beran
TIGER Programme
Institute of Hydrology
Wallingford, UK

Introduction

As preface to an analysis of education and training for Global Environmental Change (GEC) scientists I wish to pose four questions:

1. What are we teaching?
2. Who are we teaching?
3. Why are we teaching it?
4. How do we teach it?

These questions can be asked "in the round" as an aid to setting the scope of the Workshop, or they can be posed quite specifically to help to design an individual teaching programme.

The first question concerns the definition of the science taught. It is the one to which I will devote most space for two reasons:

1) a wide "between-scientist" divergence of view and
2) a "within-scientist" tendency towards an unnecessarily narrow view of what constitutes Global Environmental Change (GEC).

The other three questions bear on the relative emphasis to be given to the different components of GEC and to the style of teaching. A relatively small space is devoted to them because I am not a professional educator, but one who would have benefited from a taught course. My other qualification for entering into this field is that I conducted a small survey through the NATO region on what training facilities are already in place. My comments should therefore be seen as an amateur's *aide memoire*.

What are we teaching?

NATO's intention in mounting this workshop within its Special Programme on the Science of Global Environmental Change, was to promote a discussion on whether there is a need to teach the science of Global Environmental Change and, if there is, what should be taught and how.

NATO ASI Series, Vol. I 29
Global Environmental Change in Science:
Education and Training
Edited by David Waddington

It is rather revealing that, despite a decade or more of practical progress in the science of GEC, there is still no agreement on what is meant by Global Environmental Change. One explanation is that the GEC issue is multifaceted, and its science has been addressed through multidisciplinary teams that are often distributed through multilaboratory consortia. With all of these "multi's" it is perhaps unsurprising that there is still an absence of consensus arising from the different biases that are carried to GEC by the specialists who work in the field. True consensus will probably not be achieved until the *multi*disciplinary has been replaced by *inter*disciplinary actions.

Another contributory factor is that GEC is an issue, as well as a science. This introduces additional groups into the community of interested parties: those concerned with actions, ethics, and policies. Currently it is the "greenhouse effect" which has the public's main attention, and this has led some to simply equate Global Environmental Change with Global Climate Change. While it is certainly arguable that at the time of writing climate change is the single most worrying global scale threat it is actually but one of a set of major global environmental issues. Equating GEC with any one issue is an error that should be avoided in the workshop.

One might then equate GEC with the sum of the global environmental issues. But as well as leaving the definition unresolved, this too lacks generality, since it would logically confine attention only to those issues that have global impact - mainly those which have a negative influence on the atmosphere.

Conscious that criticism is easy, I wish to suggest another definition for Global Environmental Change. The following approach to a definition is a personal construct not found in textbooks but it is one that I have found helpful when explaining the uniqueness and the unity of the science, and useful in setting bounds to what might otherwise develop into an amorphous and arbitrary set of large-scale questions.

Two analogies help in constructing this definition of GEC:

a) Family tree
b) Venn diagram.

The family tree analogy allows us to trace a genealogical descent from Global Change, the "grandfather", down to GEC, the "grandchild". We will need to have clear a distinction between Global Change and Global Environmental Change. Traditionally the term *Global Change* has referred to a branch of earth science which deals with the "deep palaeo". To most people *Global Environmental Change* comprises a set of issues such as climate change, ozone depletion, biodiversity reduction, desertification, eutrophication and so on.

Scientists who study Global Change address major questions relating to how long the earth has had an atmosphere, or a magnetic field, or a salty ocean; and why there are ice ages, soils and mountains. Perhaps the biggest question of all for Global Change scientists is how - despite tectonic upheavals, an ever brightening sun, and bombardment by planetesimals - life persists. In attempting to answer these questions, two valuable concepts have emerged from a marriage between Global Change and a branch of engineering mathematics termed Systems Science:

(a) the earth as a system, and

(b) negative feedbacks that keep the system operational.

The combination of these two concepts is called *Earth System Science*. The characteristics of the hydro-, litho-, atmo- and bio- spheres, which make up the skeleton of the earth system, the role of biogeochemical fluxes in linking these system components, the crucial importance of negative feedbacks in maintaining the stability of the whole, and the way that the very largest (the earth) appears to be under the ultimate control of the very smallest (photons, radicals, microbes) are all Earth System Science topics of great relevance to GEC. However, the object of this article is to try to generate a shared understanding of the boundaries of the subject, not to list syllabus topics.

Returning to the family tree analogy, Earth System Science (the father science) now begets a child - Global Environmental Change science. This comes about through an alliance between Earth System Science and man-made environmental changes such as land-use changes and industrial and agricultural emissions. This relationship is possible because these man-made changes appear to attack critical points along the chain of feedbacks studied under Earth System Science - *eg* the radiation balance - and so risk destabilizing those mechanisms that have been responsible for past stability.

In the Venn diagram analogy Global Environmental Change is perceived as the intersection of two sets: one containing Global Changes (GC, *sensu* geologist), and the second containing Environmental Changes (EC, *sensu* "environmental" activist) on $GEC = GC \cap EC$.

What is helpful about this line of thought is that it avoids making special cases for changes that are less than global in impact. The latter are normally justified by reinterpreting "global" as meaning merely "big", "important", or "widely felt". Indeed most examples of GEC have to be geographically circumscribed, for example, desertification and marine pollution. Conversely, every widely experienced environmental change is not a global change, as illustrated by the following anecdote. According to contacts in the UK Environment Ministry, the two environmental problems of greatest public concern (as measured by the size of the ministerial postbag) are dog-dirt in public places and the playing of loud music. Neither can be called global on my definition (no earth system process involved), and nor would become so, either even if the whole globe shared these British preoccupations!

To express the idea briefly: *an Environmental Change becomes a Global Environmental Change if it has a significant effect on an earth system process.*

A feature common to many Earth System feedbacks is that they involve the earth's radiation budget. Industry, agriculture, land-use change, and the sheer force of our numbers have the capacity to alter the composition of the atmosphere and the properties of the earth's surface, all known points of sensitivity in the feedback systems that global change science tells us have kept the earth functioning properly for 3 to 4 thousand million years. A small intensification of the chlorine cycle and we let extra UV radiation through the atmosphere; a small nudge to the carbon cycle and we run the risk of causing climates to change; a larger nudge to the sulphur cycle and we degrade soils and hardly anybody in the northern hemisphere knows what a really clear sky looks like; a bigger nudge to the nitrogen cycle and we enrich biota but impoverish species; and with an even larger nudge to the phophorus cycle, we run the risk of doing the same to remote oceans.

I have spelt out a list of global environmental changes and the biogeochemical cycles that are implicated. It is worth reiterating that not all of the changes are global in the sense of being experienced worldwide. For example, not all places on earth consist of land that might become desert, nor are all species being impoverished because of biota enrichment. Nevertheless they are examples of GEC because they arise from interventions in the global control systems. It may be argued that this approach is no less arbitrary than either one based on whether changes are global, or on an agreed litany of big issues. I maintain that this approach does include the big issues and moreover has the considerable pedagogic advantage of permiting a variety of apparently disparate topics to be addressed inside a single framework.

Another advantage of this approach to defining GEC is that it distinguishes cause from effect. Social scientists in particular are inclined to include forcing factors like land-use change, over-population, and exploitation of natural resources in the same list as climate change and desertification. Widely felt impacts of GEC, such as increased flood risk due to sea level rise, are also sometimes bracketed with the change. For a scientific analysis it is more natural to separate the causes and the consequences from the basic phenomena.

While it is true that this scheme gives primacy to Earth System Science and biogeochemical cycles, it does not equate GEC science with Earth System Science. Only processes that operate in the human time frame are relevant. The rock cycle has controlled atmospheric and ocean properties on the multimillion year time scale but has no relevance to the current greenhouse problem. So whatever local problems they cause, mining and quarrying do not cause global environmental change. In general, biogeochemical exchanges with the biosphere and atmosphere are most relevant for GEC.

In summary I have proposed a way of thinking about GEC that links it to recognized bodies of science. It should therefore give the student the feeling that it is a science with a coherent structure. There are many ancillary topics relating to the causes and consequences of GEC that I have not addressed here. This is not because I think forcings and consequences are unimportant or that they have no place in a taught course, it is because they are somewhat independent of the way in which the core science is structured. Some thoughts on individual topics are included as Appendix A.

Who, why, and how are we teaching GEC?

Having a broad encompassing framework for GEC does not of itself determine the shape of a curriculum or the detailed contents of a syllabus. This can emerge only after we consider who is receiving the training and for what purpose. The different age groups/levels are

- senior school (school years 12 and 13 in the UK)
- freshman undergraduate
- senior undergraduate
- masters and other post-graduate
- career.

The different motivations include

- awareness of environmental issues
- learning science
- training for a research career
- training for a professional career.

There is a correlation between the two lists in that much of the interest at school level will be for more general citizenship purposes, and at the upper levels the need for information on GEC by someone in a senior position is likely to be tied to considerations of policy and advice rather than to advancing the science.

However, there are many points of overlap which weaken the correlation. Particularly important at school level will be a strong motivation for using GEC education as a means of making scientific concepts more relevant. This same motivation may well apply at freshman level where GEC science may be taught, even though it will not become the student's main subject at senior undergraduate level. The senior undergraduate teaching programmes will be beneficial to the taught element of masters and other post-graduate work, particularly in informing students more broadly beyond the confines of their research or dissertation topics.

GEC is by its nature and parentage multidisciplinary. A paradigm for higher education is that there must be discipline at its heart. I have not heard this paradigm critically analysed, but I assume it is based on the need for training in thinking, not just in providing facts. We presumably will not spend time in the workshop in re-examining accepted wisdom, but we need to acknowledge that it does give rise to a built-in tension. The obvious way to resolve this tension is through some educational system consisting of a major subject to give depth and a set of minor subjects, which convey extra GEC information. But are there other structures? Does it mean that there should never be an undergraduate degree in "Global Environment Changeology"?

In my opinion GEC requires a certain way of thinking that may be difficult to acquire unless it is taught formally. The body of facts may be no more particular than other subjects, but the need for appropriate compromise is pervasive. Is it any more than the remark attributed to Aristotle that an instructed mind should be satisfied with the requisite degree of precision demanded by the task in hand and should not seek more. In this regard GEC may be thought to resemble engineering, but the part played by principles appears much greater when dealing with the earth system. While other applied sciences can rely on empirical findings, GEC science continually needs to return to fundamentals like the periodic table, the gas laws and cellular processes.

The workshop should discuss special needs and attitudes relevant to those who will become involved in giving advice. There has been something of a paradigm shift in this area of recent years, partly brought about by some spectacular failures in the sphere of development. A story told by Joe Farman (of Ozone hole fame) comes to mind. When asked about the risk of an ozone hole over Europe, Joe said that the rate of depletion was slow and confined to the time of year when radiation was in any case low. He also said that people who lived in sunnier regions were currently more exposed to UV-B than a European would be even if the hole became worse. He suggested that the questioner could stop worrying. This drew the sharp rejoinder that it was Joe's place to give the facts; it was the questioner who would decide whether or not she would stop worrying!

This story helps make several points: the need to work with people and to give them credit for knowing their minds, and when requirement for skill in advocacy. In part such empathetic skills are innate, but much depends on having confidence in our knowledge. With such a broad subject area as GEC this does present special difficulties therefore building confidence should be one of the particular aims of GEC training. One way to achieve this is to develop a capacity to forecast an approximate result and refrain from giving extreme answers or implausible opinions. Environmental problem-solving provides many opportunities for this and course material should capitalise on these.

Teacher's perspective

I close with a few remarks concerning teachers as elicited from discussions with course providers and designers.

We need to distinguish multi- from interdisciplinarity. GEC science is much affected by both, although the latter is the ideal and should be the *aim* of training courses for scientists. Skills in team work are required for both. Teaching and learning are not the same thing, and teaching can get in the way of learning. We touch here on good pedagogical practice, which is not the particular preserve of GEC science. However the subject has the capacity to energize young trainees into exploring issues for themselves, especially if they have access to data sets and "toy" versions of models of global systems. Appendix B provides a list of centres of GEC educational interest.

Attitudes to texts and teaching resources tend to polarise members of the teaching profession. Some take the lofty view that the good student is best served by being obliged to make personal discoveries, while others seek out computer-assisted learning, video, and distance learning aids.

Higher education institutions tend not to reward cross-discipline endeavours. Structures are needed at the institutional level to assist career development of individuals who choose to work across discipline boundaries. This includes recognition and credit for joint supervision of doctoral research and publication in multidisciplinary journals.

Appendix A - Topics for taught courses

The lists below are intended as an aide memoire without special pretensions for completeness and certainly without prescription as to the depth appropriate to an individual course or student.

Basic science concepts
Physics - radiation, absorption, diffusion, transport
Chemistry - reactions, kinetics, pH, eH, "chemistry of life"
Biology - levels of organization, cell processes, dynamics
Geography - geomorphology, biogeography, human geography
Mathematics - chaos, complexity, systems theory, model building

Earth system science concepts
Reservoirs and fluxes
Conservation laws, rate laws
Feedbacks, linear and non-linear systems
Earth history - solar system, geological eras, evolution and radiation
Biosphere - plant distributions, trophic relations, exchanges
Lithosphere - erosion and weathering
Atmosphere - layers, circulation, chemistry, radiation
Hydrosphere - ocean circulation, freshwaters, phase change, transport
Cryosphere - ice caps, glaciers, radiation budget

Biogeochemistry
Weathering processes, humus formation, carbon stores
Biochemistry and cellular processes
Cycles of water, C, N, S, P, etc.

Causes of change
Industrial emissions
Agricultural emissions
Land use change, deforestation
Population and resource use

Global Environmental Changes
Greenhouse effect
Ozone depletion
Global climate change and warming
Reduction of biological diversity
Desertification and soil degradation
Eutrophication
Marine pollution
Acidification

Socio-economic consequences and responses
Sustainable development - Rio declaration, Agenda 21, minority issues
Conventions, laws - climate, biodiversity, sea, poles, desertification
Control measures - financial, legal, and educational
International policy and research structures

Tools and techniques
Modelling
Sensors and instrument systems, satellite remote sensing
Monitoring and networks
Data banks

Special skills
Advocacy
Empathy
Global thinking
Dealing with uncertainty
Interdisciplinarity and team building

Appendix B - Contacts for Global Environmental Change Teaching Initiatives

Head of Environmental Education and Training
Commission of the European Communities
DG-XI, Brussels

Education in Global Change
ICSU Committee on Teaching of Science
Professor D J Waddington
Department of Chemistry
University of York, Heslington
York YO1 5DD, UK

Education in Atmospheric Chemistry and
Global Change, IGAC Project Convenor
Professor J W Winchester
Department of Oceanography
Florida State University
Tallahassee, FL 32306-3048, USA

Dr N Azimi
UN Institute for Training and Research
Environment Training Programme
Palace des Nations
CH-1211 Geneva 10
Switzerland

Dr Donald Johnson
Dir, Earth Systems Science Education Prog
Space Science and Engineering Center
1225 West Drayton St
Madison, WI 53705, USA

Global Change Education Diskette Project
National Geophysical Data Center
NOAA, E/GC1, Dept 891
325 Broadway
Boulder, Colo 80303, USA

IGBP-START
1825 K St NW
Washington DC 20006, USA

Secretariat
European Association for Environmental
Management Education
viale Ippodromo, 9
I-21100 Varese, Italy

Institut Laplace
10-12 Ave de l'Europe
78140 Vélizy Villacoublay
France

Education and Communication on
Environment and Development (ECO-ED)
720 Bathurst St, Suite 500
Toronto M5S 2R4
Canada

Dr D Troost
Training and Education in Marine Sciences
Unesco
1 rue Miollis
75732 Paris Cedex 15, France

Vice Rector, United Nations University
15-1 Shibuya 2 chome
Shibuya-ku
Tokyo 150
Japan

The Director
Stockholm Environment Institute
Box 2142
S-103 14
Stockholm
Sweden

Education and Training Division
World Meteorological Organization
Geneva 20
Switzerland

Dr J J Nijman
EERO Training Centre
PO Box 182
6700 AD Wageningen
Netherlands

The Editor
Journal of Environmental Education Research
Carfax Publishing Co
PO Box 25
Abingdon OX14 3UE, UK

Carbon Dioxide Information Analysis Center
Oak Ridge National Laboratory
PO Box 2008
Oak Ridge, Tenn 37831-6335, USA

Int Environmental Education Programme
Unesco
7 Place de Fontenoy
75700 Paris
France

Sustainable Development
Towards Global Change partnerships

W S Fyfe
Deptartment of Earth Sciences
University of Western Ontario
London, Ontario
Canada, N6A 5B7

Introduction

The twentieth century has been truly remarkable in the development of natural science. The giants of the early 1900s laid the foundations for the description of matter and energy, and the developments of the past few decades have led to our present age of observation. Given support we can "see" almost everything from a galaxy to a genetic defect. The same developments have led to the age of information and modern computers. And via the new communication systems, particularly television, we are in the age of expectation. Television, which influences the daily life of almost every human of every age, shows more and more the differences between rich and poor - those who starve to death and those who drive a Buick. Today, about 40 nations have deep internal social conflict. In the 1940's the great British writer, Aldous Huxley said we were driven by a creed of greed, violence, and incomprehension. We still worship wealth and violence, but at least the incomprehension is no longer excusable.

The science and resulting technologies of this century have supported the phenomenal growth of human population. In 1750, less than one billion; 1850, about 1.2 billion; 1950, 2.5 billion; now we add almost 100 million per year, and the World Resources Institute in their brilliant reports (1994) project over 10 billion by 2050. Population in Africa grows annually by 2.93%; Asia, 1.78%; the world, 1.68%.

The technologies which have supported this phenomenal growth are changing the most basic elements of our life support system. We have perturbed the chemistry of the atmosphere and the emission of gases such as carbon dioxide is increasing. Soil degradation is catastrophic. As Postal (1992) shows, at least forty nations have an acute shortage of water. There are clear signs of unusual climatic conditions, record temperatures, floods.....

NATO ASI Series, Vol. I 29
Global Environmental Change in Science:
Education and Training
Edited by David Waddington

King and Schneider (1991) of the Club of Rome stress that in many of our strategies for development, for the use of knowledge, we have failed. In large part this failure can be attributed to the recent systems promoting specialization. It is interesting to reflect on how this has arisen. First, industry needs specialists, the best material for a semi-conductor, the best alloy for a jet engine....the best chemical to kill a mosquito.... The agencies which fund research look for the specialist, and they support each other. The ultimate insult is to tell the scientist that his work lacks focus! And in this century a vast amount of the support for basic research has been derived from the need to have the best weapons.

Further, the demand for specialists for our modern technologies has led to a huge expansion of the university systems. Early this century universities were small. The student of chemistry or physics probably knew the history professor and drank coffee with students from the arts. Today, this is rare. The great concern of the modern student is, "Can I compete to obtain a job?" University departments and their students have retreated into their cocoons of specialization in order to survive and compete.

Thus, while there have been amazing advances, the global situation is causing vast concerns. The Stockholm (1972) and Rio (1992) Earth Summits demonstrate these concerns, but they have been failures in terms of effective action.

The ultimate objective of science must be to improve the human condition for all people. Since the famous report of the World Commission on the Environment and Development (the Bruntland Report, 1987), a term increasingly used is 'sustainable' development. Many argue what this means. For me it is simple: will I leave the planet in better condition for all, all species who support me, and who will follow me? It is unfortunate that the Bruntland report is very anthropocentric, and lip service only is paid to the crucial question for the importance of biodiversity.

The Future

To support well the growing human population, there will be, there must be, vast development. Urgent problems are concerned with areas such as:

- energy for 10 billion
- water supply
- soil conservation
- transport systems

- waste disposal
- mineral resources

All on a planet where the 'environment' never was, never will be, in a steady state (e.g., Stine, 1994). This implies that for security in the future there must be reserves and technology to distribute such reserves.

In the past, strategies for development have come mainly from specialists with the support of economists. I agree with King and Schneider that many of these developments have been truly non-sustainable. It is clear that if the situation is to improve, we require new systems. I remember well the words of the late Roger Revelle, the director of the Scripps Institute of Oceanography. He said the needs were "urgent and hard". Urgent for obvious reasons. Hard, because they involve cooperation between the specialists and across disciplines.

The Need for Teams

Environmental problems are complex. The earth environmental system and its mass-energy exchanges involve interactions between:

- the sun
- the ocean-atmosphere transport systems
- the biosphere
- the geospheres
- and now anthropogenic technologies

These all fluctuate on a wide range of time scales from seconds to millennia.

It requires specialists to study any aspect of these complex forcing interactions. I don't think that many of us have the time or ability to be an expert simultaneously in solar physics, the response of a plant to changes in atmospheric carbon dioxide, and the impact of a new defoliant on the albedo. This implies that to arrive at the best systems for the needed new developments requires development of improved systems for effective cooperation.

I very much like the words of Jay Forrester, a systems analyst quoted by King and Schneider: "In any complex system, attack, however apparently intelligent, on a single element or symptom generally leads to a deterioration of the system as a whole". Examples abound: If we pump groundwater for agriculture at a rate greater than replacement, the system will crash. If we burn all fossil carbon; i.e., oil, gas, and coal, as rapidly as possible, we will have a greenhouse.

In terms of needed developments and the teams needed, every case will have very specific needs. There is no doubt that our standard of living is related to the availability of energy, and today over 90% of the world's energy comes from oil, gas, and coal. There is also no doubt that standards of education are related to energy, and in those fortunate lands where there is universal access to quality education, the population problem is coming under control (see data in World Resources 1994). For example energy consumption per capita (1991, giga-joules) is 320 - U.S.A., 187 - Germany, 9 - India, 12 - Africa, 23 - China. While some waste, most have great need.

Imagine that we faced the challenge of providing increasing availability of energy for the developing world (I hate that term!) by a factor of ten. Who would be on our team? For certain we must consider the sustainable energy alternatives:

- solar of all types (photoelectric, wind, biomass......)
- geothermal potential
- nuclear potential
- fossil fuel potential

We would need experts in all aspects of food-fibre production, water resources, solar technologies, climate experts and statisticians, biodiversity experts, nuclear technologists......... And teams, local and global, must be involved in the planning *from the start* and be proactive not reactive.

The Technological Opportunity

I think that there is little doubt. Those who recognize the great environmental problems and act, will lead the technological structures of the next century. The crisis areas, some of which are listed above, all require new systems on a grand scale. There is a vast opportunity for the innovative practical mind. The specialists we train in universities must be kept in contact with the environmentalists, industrialists, and economists. But as always, there are opportunities at more simple technological levels (e.g., low tillage agriculture for soil conservation, or the work we are doing in Canada and India, using sewage and coal ash to greatly enhance the growth rate of trees and remineralize degraded soils). For a vast array of our present environmental waste problems, there are simple solutions of easily proven economic advantage. The key to action is to assemble the groups involved and engage them in realistic dialogue.

The Role of Education

If we are to improve the present situation and systems, we must also improve our educational systems that ultimately drive our political systems. I very much like the words increasingly used: "an objective of universal literacy, numeracy, and science". It is strange that on our planet so many do not know why we have so much oxygen in our atmosphere, that if oceans are warmer there will be more rain somewhere, or that frogs can control mosquitoes. The average citizen has no idea what happens to waste products. Planetary science has something for every student of every age, of every ability. The challenge is to choose the best examples to illustrate general principles. I have had great success with elementary school children in discussions of the economics of saving energy and stopping waste production in general. Children are fascinated to learn that ECO-logy and ECO-nomy have a common Greek root, Oikos-house, no conflict!

At university all students must have time to learn a little about the general topic of the Earth System and their life support systems. For example, I think all students should read the reports of organizations like the U.N. population council, the Worldwatch Institute, and the World Resources Institute which includes wonderful data tables (also available in diskette form). But perhaps most of all we must develop systems that allow students from different disciplines to meet, discuss, and debate, formally and informally. We must have specialists, but we need specialists who can communicate with other specialists and who can overcome the jargon of their specializations. A great advantage we also have at the university level is that normally there is a significant range of national and ethnic diversity.

We have always been concerned with the problems of illiteracy. I suggest that today the most serious problems we face result from the total lack of knowledge, ignorance about why we are here and how we have evolved. This ignorance is vast, from the Pope to our leading politicians.

At this university we have started an experiment. At the graduate level a student of astronomy, physics, chemistry, and so on, whose work has application to the development of the environment, can elect to graduate Ph.D., M.Sc., Discipline - Environment. All these students must attend a full course on the planetary environment - from solar physics to convection on Earth. This course is given by faculty from about 12 disciplines. Then all students must attend their own course, where they present the topic of their research and describe why it is important for the improvement of the environment. It is a start, but we have a long way to go.

We need more specialists. All must read and write, add and subtract. But all must learn to respect their ultimate home. And hopefully, whether by positive action or catastrophe, the population is reduced, and the billion or so humans remaining will have a joyous planetary experience.

I like the words from a green U.S. company, Ben and Jerry's Ice Cream: "What we cannot do alone, we can do together".

Or is Richard Dawkins (1991) correct: "Like every plant, ant, worm, we are simply bio-robots."

References

Dawkins R (1991). A new world view, in, from Gaia to selfish genes. Ed., C Barlow, MIT press, Cambridge, Mass 195-212.

King A and Schneider B (1991). The first global revolution. Simon and Schuster, London, 397 pp.

Postal S (1992). The last oasis-facing water scarcity. WW Norton and Co. New York, 239 pp.

Stine S (1994). Extreme and persistent drought in California and Patagonia during mediaeval time. Nature 369: 54 6-549.

World Commission on Environment and Development (The Bruntland Report) (1987). Our common future. Oxford University Press, 383 pp.

World Resources Institute. (1994). World Resources 1994-95, a guide to the global environment. Oxford University Press, 400 pp.

Environmental education and training of global scientists for sustainability

J R E Harger
UNESCO/ROSTSEA
Jln. M.H. Thamrin 14
Jakarta 10012, Indonesia

Dirk Troost
UNESCO
1, rue Miollis
75732 Paris, France

Introduction

An element in modern thinking concerning environmental education as defined by "environmental educators" concerns the notion that one may divide the field into: 1) education about the environment and 2) education for the environment. It is proposed by some that "environmental education" is not "environmental science education". This seems to mean that environmental education (education for the environment) is exactly what one might call "environmental indoctrination" or even "environmental ideology". Although this may be acceptable to some extent, it seems to be entirely removed from the environment and apparently deals with 'values', and 'ethics'. However, the way in which these values are arrived at is not stated. Likewise the problem of rationalizing an ethic is simply avoided. In using this approach, nobody gets to go into the environment to actually check up on the real situation, since all answers are apparently known, and the idea that students should be encouraged to verify information for themselves is not well supported.

It is important to stress that "environmental-global change education" should start as early as possible within the education experience, and this should take place at least in high school. This form of education concerns assessment of the real world with the intent of using the resulting information to obtain 'reality checks'. It involves transmission of the viewpoint that calls everything into question and seeks answers, concerning the processes governing sustainability of the biosphere, from the earth. It deals with the methodologies involved in testing and verifying elements of local, regional, and global environmental change. Values and so forth flow from the information deduced and adduced in relation to the question of "environmentally sound and sustainable development". The element to consider in all contexts is sustainability. It is, however, important to ask what exactly this term means in the context of scientific practice, with respect to the information that science brings to light, in respect to social practice, in terms of environmental processes, and so forth. The

NATO ASI Series, Vol. I 29
Global Environmental Change in Science:
Education and Training
Edited by David Waddington

term "development" should not be regarded as daunting since it can at the limit refer to qualitative factors. This approach can provide a realistic basis for elaboration and definition of the values the educators are looking for. The practices and attitudes that are of value to society are thus those which are in line with sustainability.

The question underlying local, regional, and global sustainability and its definition and analytical relation to all that is learned about the environment should be the mainstay of all training for environmental global change research and understanding. This is important because comprehension of the term "sustainable", in the context of the principles governing maintenance of the biosphere, actually provides the basis for values and ethics in the modern world. Behavior that promotes sustainability, or at least that does not promote the opposite trend, should be regarded in a positive light and seen as ethical. Positive value thus would lie in the reinforcement of sustainable practice in all respects. The mode by which science interacts with this form of analysis is thus of some importance, since it is scientific practice that can be used to frame, analyze, and test the validity of claims that will be made towards sustainability in everyday affairs.

This notion is most important because presently there are rather significant distortions of the work carried out under the Global Environment Facility (GEF) research programs, that in some ways amount to focal points on one side or the other of the north-south (developed-undeveloped) split. For instance, work on methane production associated with rice growing appears to be undertaken in an effort to actually show that the developing world is at fault in generating significant contributions to greenhouse gas sources as the result of depending on rice as a staple. Nevermind that most rice fields are now located on what were previously methane-producing wetlands, that rice merely vents methane that has formed from the previous year's green fertilizer, that the carbon released goes up mainly as CO_2 and will be reclaimed from the atmosphere as rice growth proceeds, that the slightly positive methane/CO_2 ratio that might develop is a small fraction of the carbon/global warming term incurred by the fossil fuel based agricultural effort, and so forth.

Field work philosophy

The educational programs required to combat environmental change must also be anticipated in the form of concretely focused studies for graduating students.

Problems associated with environmental management options will increasingly require the attention of an educated population to render solutions compatible with the emerging environmental ethic. Educational institutions will face new pressures to provide adequate training in this regard, and the best way of ensuring an appropriate response is to expose students to field situations requiring their attention to observation and detail at an early stage. Indeed, examples drawn from environmental studies can be used to illustrate almost all facets of a modern scientific education, and the increasing use of the phrase "environmental education across the curriculum" is an indication of the extent to which serious environmental studies are now required at all educational levels and, most importantly, in high schools.

There is widespread agreement that the principle of *ecologically* or *environmentally sound and sustainable development* should be adhered to in both developed and developing countries. An important step in ensuring that this will be undertaken on a broad front involves engaging high school students with the realities of environmental analysis, so they may contribute to defining exactly what future generations will mean by this (at present) uncertain phrase. High school science faculties will increasingly be called upon to design interlocking programs around the major areas of environmental studies, although at first this may only involve token adjustments, such as the inclusion in chemistry courses of behavioral characteristics of common salts in solution to promote understanding of saline versus fresh water environments and so forth. Yet, at a later time it may be possible to imagine a far wider array of interlocking subject areas within standard and classical curricula, which will be pointed towards an increasingly sound and scientific understanding of environmental processes, and eventually lead to the development of specific curricula on the subject.

Field studies objectives

Properly planned field studies can provide the opportunity for team collection of environmental data at sufficient densities so as to require the use of microcomputers for analysis and reporting. These same data sets over the long term may then be saved to provide a systematic time sequence for the detection of secular change. Field studies should form a key part of an interdisciplinary curriculum that can prepare present and future generations with an understanding of marine, terrestrial, atmospheric, and associated systems, as they relate to ecologically sound and sustainable development in a continually changing world. A field studies curriculum should bring together elements within the existing syllabus to: (1) make advanced knowledge of marine terrestrial and associated systems accessible to teachers and students through the analysis of existing problems; (2) catalyze interaction among high schools and established research and education systems to ensure that activities significantly contribute towards required educational goals and are accepted as credits towards qualification in different programs where appropriate; (3) encourage the collection, recording, and dissemination of information relating to the environment as the result of the accumulation of class data over the years; (4) encourage implementation of standardized monitoring and testing protocols to enable consistent utilization of data over a wide front; (5) progressively establish educated communities in all countries of the world to understand factors associated with environmental change and the resultant impacts, thereby encouraging global ecologically sound and sustainable development in a changing world; (6) at a broad level encourage sharing of knowledge, resources, and facilities, and selectively enhance facilities to maximize the capability of all participating institutions.

Ecologically or environmentally sound and sustainable development

A major concern in environmental thinking and practice currently deals with the area known as "environmentally sound and sustainable development". Implementation of well thought-out national ecologically sound and sustainable management plans would be a major asset in the development of an overall response to environmental change. A systematic approach to this problem will permit the allocation of resources to upgrade public analytical thinking about the oceans and the terrestrial environment and their role in global change. It will also contribute to significantly raise the level of awareness for environmental problems in all segments of the population at both national and global levels. Recent attempts have been made by many international organizations, including the UN system, to direct the global community in adoption of a policy of environmentally sound and sustainable development (ESSD). In a crude sense, we may define ESSD as development which permits further development without terminal closure of favorable options. In pursuit of this end, the general approach towards defining the potential for progression of civilization throughout the world must be addressed through ecologically sound and sustainable development. This in turn means that only human actions which permit continued development, with the environment as the final arbitrator, can be considered as constituting an acceptable strategy for social activity at all levels. The challenge at this stage is to elaborate useful models which will permit specification of such action. We know what we should like the phrase "ecologically sound and sustainable development" to mean, but we currently have no clear idea as to how this can be attained in practice, nor indeed how it can be measured objectively.

A major task for environmental education will be to lay the groundwork to permit definition of what a community will understand by this phrase through early experience with the subject matter at a sufficiently detailed level. Eventually this will require an increasingly heavy commitment to gather and process environmental statistics and to define and quantitatively evaluate development options. Suitably educated students will be able to participate in this exercise by contributing informed opinion, as required in the future. One of the most difficult areas to deal with will be the assessment of the effects of energy injections. On the one hand, energy enrichment to human social systems is clearly used to promote increases in carrying capacity and the elaboration of useful structures. On the other, the act of energy capture and subsequent energy degradation promotes vast and as yet unquantified negative environmental impacts and instabilities, among which the increasing concentration of atmospheric CO_2 stands out as a significant warning signal.

Organization and structure

The field exercises should be proposed to highlight and analyze environmental problems relating to the ramifications of change. Although it can be anticipated that not all influences associated with global change will be deleterious, the exercises should focus on raising student awareness of the environmental issues faced by modern society. The suggested method is to take a particular local problem or a set of problems and to analyze these in detail in order to (1) make students familiar with the particular circumstances through their own efforts (data collecting and analysis) and (2) help them gain an understanding which will relate to the whole globe.

Lack of understanding about the influences operating in the modern environment continues to be widespread. Societies in many countries have evolved through substantial periods of dependence on economies operating under apparently balanced ecological conditions. For many, the effects induced by global industrialization and global economic processes are new phenomena, ones which often cannot be understood by the direct transfer of practices which were formerly adequate to deal with local causes and effects. In the face of advances in technology and changes of scale owing to the operation of the global economy, traditional practices, such as those involved in fishing and ocean management, often prove inadequate to guide further development. For instance, in the Pacific Ocean traditional management practices have evolved successfully over long periods of time to ensure that local stocks of living resources are not fished out. The use of modern fossil-fuel-based enterprises which key into the global economy however, often result in severe over-fishing, as efforts are made to capitalize on local resources and to repay loans for fossil-fuel-powered equipment.

Environmental studies and traditional approaches

In order to be realistic, the interdisciplinary nature of environmental studies does not have to be emphasized to the extent that it cannot be accommodated into existing curricula. On the contrary, topics selected should be relevant to existing education requirements and should be related to the socio-economic needs of the community that will receive the graduating student. Any subject taught using a field-work approach should have an existing knowledge base and a theoretical underpinning. If the subject is to be used to generate effective environmental insight, it must also be taught using a practical approach involving learning by doing, problem solving, and decision making. An important aim should not be so much to orient the subject for merely passing exams, important as that may be, but to conduct it in such a way as to contribute to the fulfillment of the needs of learners in terms of generating insight into the current environmental problem and enabling those concerned to realize their own role in the issue. In other words, the educational implications of environmental studies, particularly environmental field work, as part of the school education experience must be clearly understood and subscribed to before field studies are undertaken. It is the purpose of this paper to advocate a field work experience focused in such a way that the major objectives within existing curricula can be achieved more effectively than would be the case if no field experience was offered. In addition, the information gained should be of direct use in the construction of strategies for obtaining successful examination results.

Environmental ethics

As human population pressures increase throughout the world and the influence of industrialization spreads from formerly restricted areas in ever widening circles powered by fossil fuel energy accumulated from ancient ecosystems, the factors mitigating against continuity of undisturbed natural communities capable of maintaining a buffered environment intensify. With the possible exception of remote communities in the abyssal depths of the oceans or perhaps under the Antarctic and Arctic ice-caps, many ecologists now feel that there are few, if any, undisturbed ecological communities remaining on the face of the earth.

Traditionally, ecological field-trips have been marked by a heavy emphasis on sample collection and preservation for later identification in a laboratory setting. For the most part such activities encouraged the destruction of organisms in the name of science with the majority of the formerly living animals and plants left to rot in forgotten corners. In some places on the shores of continental North America successive class field trips caused discernible damage. Even today an active search is maintained for comparatively rare shells, particularly cowries which are already considered to be under threat. The modern wilderness and eco-tourism slogan of "take only pictures, leave only footprints" has surely found its way into the consciousness of most high school students, so that it is important to design environmental studies programs having the minimum of impact upon the communities studied.

As with the Jains of India who believe that the conscious or even the unconscious harm exerted on the least of living creation detracts from their own quest for perfection to the extent that they wear no leather, silk, or even fibers obtained from living plants, preferring instead to go naked in the world and to studiously brush aside insects that might otherwise be harmed by their passage; so too should the execution of environmental field studies be conducted with the utmost in circumspection with respect to the object of study. Insofar as it is possible, the activities planned in the program developed within this context should be designed to exhibit the minimum of impact. Experimental actions against larger and rarer life forms, that result in destruction of the individuals concerned, are expressly frowned upon.

In spite of this injunction it nevertheless remains true that some form of impact will be involved with even the most carefully planned field course. Where coral reefs are examined using snorkeling techniques, stray fins will surely break some of the most fragile forms, but even here, careful preparation can hold the damage to a level that is far below that which would accompany a heavy storm. In wading across a reef flat it is certain that some organisms will be adversely affected. Yet again the teacher should devote time and attention to these details for it is only by arguing for the good of the whole by virtue of knowledge gained that the sacrifice of the few can be justified. Clearly some ecosystems such as coral reefs adjacent to urban population centers or in association with massive tourist developments are more at risk than, say, the plankton blooms that develop in response to sewage discharges (nutrient enrichment) around the outskirts of the ever growing human city ecosystems. Considerations such as these can be used to develop guidelines concerning the extent to which sampling mortality in the name of ecosystem knowledge can be justified.

In other words, the whole concept of environmental field study programs must be exercised in an environmentally sound and sustainable manner. We may approach this problem best through consideration of the extent by which it may be possible to conceive a form of natural self-sacrifice, which nature may be prepared to offer in exchange for providing the basis for the development of human understanding that will permit elaboration of survival strategies for the biosphere as a whole and thus for humankind as well. By this viewpoint, humanity depends on nature, and at this stage in the evolution of our planet, nature depends, as never before, on humanity.

Expected output and accomplishments

Improvement in the quality and quantity of environmental education and related analytical capabilities of graduating high school students will enable institutions such as universities, government agencies, and private firms to carry out their missions more effectively by having access to informed personnel. Although the final outcome of a course in environmental studies should lead to an integrated comprehension of the major factors shaping the world and the extent to which mankind may interfere with them, nonetheless this overall perspective will normally have to be reached through a variety of course options that may be offered at a particular high school. The broad questions set in examinations often require students to interpret local conditions in the formulation of answers. Often a considerable array of material can be assessed if students have been exposed to a concentrated field program where they have been forced to analyze information they have themselves collected. Questions relating to the implementation of particular sampling protocols and to ecological principals such as niche theory, the pyramid of numbers, etc., can be covered in a meaningful way in an environmental field studies program. Other general biological principles may also be addressed, including identification and classification, food-webs and food-chains, energy flow patterns, and population studies. In certain instances, the study of management practices may be considered.

The aim of the field trip is to provide an integrated exposure to environmental analysis, to enable the students to obtain unequivocal evidence concerning the effect of at least one or more environmental problems, and at the same time to provide the necessary exposure to practical examples of facts, processes, and interactions that are specified within existing syllabus plans. Since the fundamental rationale for the activity involves the use of environmental problem areas as the skeleton upon which the individual syllabus concerns are to be related, and since time and costs are limiting factors, it is essential to select a program which will accomplish both objectives.

Report production

The field experience and the associated data collection exercises form the main component of the overall activity being designed, in part to force attention of students towards structures under study. The main part of the learning component is, however, focused on the reflection required to produce a comprehensive written report. In addition to the simple description of the exercises involved and the overall account of the activity, the report should contain a detailed analysis of the data collected. The activities should be designed for implementation by study groups involving from 4-6 students (not more). The amount of data generated from a class of 30-40 can be quite significant. For instance, a coral reef transect exercise in the tropical marine environment can readily generate 12 transect readings for each student group. Each of the transect readings may involve positional information for the life-forms encountered, as well as density and coverage measurements. This form and density of information can only be sensibly analyzed with the help of a microcomputer and in addition can form the basis for studies of secular change.

An Environmental Education Framework Based on Ecological Sciences

Within the concept of "Environmentally Sound and Sustainable Development", an important overall issue is development without excessive outside interference. Relevant specific issues are (1) loss of biodiversity, (2) land degradation, (3) degraded ecosystems, (4) socio-cultural gaps, (5) socio-economic imbalance and problems associated with ensuring social justice, (6) conflict between traditional values and foreign life-styles, and (7) use of biodiversity. The issues underlying the individual geographical components of an environmental education system are, (1) requirements for the development of ESSD plans, (2) preservation and maintenance of biodiversity, (3) control of global warming through the requirement for the elaboration of carbon sinks (coral reefs, inland and mangrove forests, etc.), (4) assessment of external pressure in the form of exploitation of mineral, timber, and other natural resources, as well as eco- and cultural tourism etc., (5) relationships with national parks, (6) requirements for ecological repatriation of damaged ecosystems, and (7) requirement for development of effective local area management.

A preliminary approach towards customized courses to address environment and sustainable development might be envisaged to cover the following 3 areas:

Area I	**Area II**	**Area III**
SOCIO-ECONOMIC DEVELOPMENT	**BIOSPHERE PROCESSES AND BIODIVERSITY**	**GLOBAL CHANGE**
1. Development models/theories	1. Evolution of species, ecosystems, and biomes; function of biodiversity	1. Pollution, pollution as externality of socio-economic development
2. Indicators for ESSD	2. Ecosystem function in relation to global climate stability	2. Greenhouse warming, ozone depletion and other manmade environmental changes
3. Development agencies and activities (Stockholm, Bruntland, UNCED, post-UNCED Agenda 21)	3. Rates/causes of ecosystem changes	3. Global change and options for future
4. Natural resource accounting	4. Conservation of nature: notations from different cultural backgrounds	
5. Environmental impact assessment and cost/benefit analysis	5. Values associated with protected areas (PA's) as bio-diversity centers, conservation, management, etc.	
6. International relations	6. Protected area management and development linkages	
	7. PA's and Post-UNCED, Agenda 21	

As an example, the following outline covers the major features to be considered in an ecology course, which are also the appropriate elements of environmental studies. Particular field exercises can be slotted into the framework and given the emphasis necessary to address traditional subject areas. It should be noted that unless some thought is given to this, teachers may have some difficulty in determining the relevance of specific exercises that may suggest themselves when facing field situations.

Structural outline for ecology and environmental studies

1. The natural environment and interactions among organisms
    - 1.1 Living space
        - 1.1.1 Effects of physical factors (determining where organisms can live)
        - 1.1.2 Effects of biological interactions (determining where organisms may live)
            - 1.1.2.1 Effects of the same kinds of organisms (intraspecific competition)
            - 1.1.2.2 Effects of other kinds of organisms (interspecific competition), predation
            - 1.1.2.4 Evolutionary adjustments: mutualism, commensalism, etc.
    - 1.2 Mortality factors, physical, biological
2. Biological processes and the physical environment
    - 2.1 Role of energy in community maintenance, photosynthesis (primary production)
        - 2.1.1 Active transfer of biologically fixed energy, conditions governing stability
            - 2.1.1.1 Predation as an energy transfer device
            - 2.1.1.2 Food webs
            - 2.1.1.3 Pyramid of numbers
    - 2.2 Role of nutrients and essential elements in community maintenance
        - 2.2.1 Identification of major elements and nutrient components
        - 2.2.2 Origins
        - 2.2.3 Recycling
    - 2.3 Energy-mediated injections into natural systems
        - 2.3.1 Nutrients
        - 2.3.2 Elements (carbon, etc.)
        - 2.3.3 Complex chemicals (chloroflurocarbons, organic pesticides, etc.)
3. The biosphere
    - 3.1 Conditions associated with sustainability
    - 3.2 The Gia hypothesis, energy use, feedback control
    - 3.3 Biodiversity and long-term changes
4. Human impact on the environment
    - 4.1 Assessment of biological communities
    - 4.2 Monitoring of physical components (pollution)
    - 4.3 Reconstruction of damaged ecosystems: why do it ?
    - 4.4 Human population: growth and impact
    - 4.5 Human ecology and the concept of sustainability

The Role of Environmental Education in the Training of Scientists on Global Environmental Changes in Developing Countries

W Leal Filho
European Research Centre on Environmental Education and Development
University of Bradford
West Yorkshire BD7 1DP, UK

Introduction

Although over half a century has passed since Haeckel launched the concept of ecology, ecosystems, and environmental balance, only in the last two decades has the problem of global environmental changes become an issue to be seriously discussed. New scientific evidence shows that there is indeed a strong sign that various alterations (for example in the planet's climate) are taking place, which are bound to lead to major changes in habitats and in the ecological dynamics of various plant and animal species, as well as in human behaviour.

A number of on-going research programmes have been used to try to understand the origins, structure, and dynamics of global environmental changes. These vary from research aimed at detecting increases in the temperatures of various parts of the planet (which may possibly be linked to the so-called 'greenhouse effect'), to the preparation of models for predicting the evolution and eventual advancement of desert areas in regions such as northern Africa. A fact that is clear from the body of research already existent, and that has been registered by UNDP (1991) and IUCN (1991) in two of the most significant publications in the field of environment and development in recent years, is that the current models of economic growth, adopted by the various countries which now suffer intensively from the effect of climatic variations, are not the most suitable ones. In addition, it is clear that alternative ways of improving a country's economic performance should be found that are preferably based on the principle of sustainable development.

However, due to the magnitude of the problem and its various variables, many research topics are yet to be covered comprehensively. Leal Filho (1992) listed some research needs seen in terms of global environmental changes in developing nations, with a special emphasis on climate changes, and concluded that many items are yet to be fully investigated.

NATO ASI Series, Vol. I 29
Global Environmental Change in Science:
Education and Training
Edited by David Waddington

Among today's research needs, mention also ought to be made of investigations of the ways through which national policies implemented in developing nations tackle the problem of global environmental changes. Such a need, as acknowledged by Magalhaes and Glantz (1992), is illustrated by the fact that, although various studies have dealt with policy on global environmental changes in industrialized nations, there are few registers of comparative studies which have focused on the political behaviour and the social responses from developing nations to these changes. This is especially conspicuous in economically powerful developing countries, such as Mexico and Brazil, in countries considered overpopulated like India, and in countries which severely suffer the effects of global changes, such as Bangladesh and the Philippines.

From Discussion to Action

Despite the research needs discussed above, a steady evolution has been seen in the process of discussion on the item "global environmental changes". For example, being aware of the increasing evidence of climatic variations, the World Meteorological Organization (WMO) set up its International Climatic Programme in 1976. This development was followed by a conference in 1979 about the world's climate, the World Climate Conference. In that event, the concept of global warming, or the greenhouse effect and the role of CO_2 in exacerbating it, was one of the most discussed items. A few years later in 1985 a conference in Villach, Austria, approved a set of research studies on CO_2 (by then clearly acknowledged as a major factor associated with climatic variations), which included quantitative and qualitative analysis of CO_2 concentration in the atmosphere, and its effects on the climate and on ecosystems. The Villach Conference set in motion further research studies.

More recently, the United Nations General Assembly at its 1990 session set up the Intergovernmental Negotiating Committee for a Framework Convention on Climate Change (INC/FCCC), which was given a mandate to "draft a framework convention and any related legal instruments it considered necessary" (IUCC 1993). Since then, negotiators from all over the world have been working on the text of such a convention. In May 1992, they adopted the United Nations Framework Convention on Climate Changes which was signed by more than 150 countries.

According to the latest ratification status report, over 50 countries ratified the convention and a meeting of all ratifying countries, dubbed as the "Climate

Summit" is expected to be held in Berlin in March 1995 to discuss its implementation. Once this process is completed, the INC/FCCC will be disbanded and the responsibility of the fulfillment of the convention's recommendations and implementation will rest with the Conference of the Parties, a body to be created, which will be composed of representatives of the countries which ratified the convention.

The Need for Training

In consideration of the need for education, training, and awareness as tools for the implementation of the convention and as instruments for effectively dealing with global environmental changes as a whole, the United Nations Framework Convention on Climate Changes has made provisions for their incorporation as part of its structure.

Article 6 of the convention (Education, Training, and Public Awareness) requires the parties to promote and facilitate at the national, sub-regional, and regional levels the following measures:

1. the development and implementation of educational and public awareness programmes on climate change and its effects
2. public access to information on climate changes and its effects
3. public participation in addressing climate change and its effects, and in developing adequate responses
4. training of scientific, technical, and managerial personnel
5. the development and exchange of educational and public awareness materials on climate change
6. the development and implementation of education and training programmes, including the strengthening of national institutions and the exchange or secondment of personnel to train experts in this field, in particular for developing countries

Due to the relevance of training especially designed to correct the deficiencies in the knowledge of the relationship between environmental policies and environmental changes (at the local and global level) found in developing nations, some considerations of the role of environmental education, as a tool for understanding global environmental change as a whole and climate change in particular, should be made. Environmental education, as a process aimed at raising awareness on environmental issues (Leal Filho 1993, Unesco 1987) may catalyse action aimed at reversing the lack of information seen today among people in developing nations, especially among policy-makers.

The systematic provision of information on environmental matters, one of the applications of environmental education, is also very useful in the sense that the body of knowledge gathered from training may be used as reference points in helping developing countries to prepare more reliable national policies, which would allow them to properly deal with global environmental changes.

An analysis of some on-going national environmental policies related to global environmental changes was undertaken using a sample of six developing countries. In the analysis, the sample countries' effectiveness under pre-determined criteria was evaluated, and a comparison of the basic lines of their policies was made with a view of suggesting ways of improvements. The study, which consisted of a survey involving a sample of 6 developing countries and done over the period 1992-1993, aimed to:

- investigate the existence of policies associated with global environmental changes in individual countries
- assess the effectiveness of the existing policies in individual countries on the basis of their responses to environmental problems (including disasters) and on their preparedness to deal with global environmental changes
- suggest, based on the principles of environmental education, some items that should be considered in the setting up and/or review of national environmental policies, aimed at improving a country's likelihood of effectively dealing with sudden environmental and climatic variations and handling long-term environmental changes

In terms of the sample used, two representatives of the three largest geographical areas of the planet, where developing nations are concentrated, were chosen. In Latin America, Brazil and Mexico were selected on the basis of their size and economic influence in the region. In Africa, focus was given to Morocco and Kenya, the former having a large desert and the latter suffering an advanced process of desertification, which makes an investigation into their level of response to global environmental change a promising issue. In terms of Asia, Bangladesh was selected due to the fact that it is one of the region's most populated countries and because it has been victimized by various environmental problems related to global changes. The second Asian country, namely the Philippines, was chosen on the basis of its historical vulnerability to environmental disasters, which makes it an interesting case study for assessing the existence and scope of policies oriented towards environmental variations and long-term environmental changes, adopted by a developing country. Language barriers were probably minimal due to the investigator's fluency in English, Portuguese, French, and Spanish. Data collection consisted of a combination of one empirical and two theoretical methods. In terms of the theoretical methods, a historical investigation of the scientific evidence for environmental changes in the selected countries was undertaken. This was combined with a structural analysis of national environmental policies. In empirical terms, an evaluation of the countries' infra-structure for dealing with the local effects of global environmental changes was undertaken. The establishment of relevant collaboration schemes with persons and institutions, nationally and internationally,.was paramount in the execution of the survey. Table 1 shows the criteria used in the survey and Table 2 shows the results obtained.

Table 1 - Criteria used in the Study

Assessing National Environmental Policies, with a Special Reference to Global Environmental Changes, in Developing Countries

a) preliminary procedures:
- diagnostic of a country's economic and political situation
- main environmental resources found and commercialized
- significance of the trade of environmental resources to the economy
- social problems related to the use of environmental resources

b) policy study

1. existence of National Environmental Policy (NEP)
2. date of implementation of NEP
3. organization(s) responsible for the NEP
4. integration between organizations
5. existence of training programmes
6. status of personnel (i.e., well trained, reasonably trained, poorly trained)
7. prospects for improvements

Table 2 - Emphasis given to Training on Global Environmental Changes in the Context of National Environmental Policies in 6 Developing Countries

Country	National Environmental Policy	Emphasis on training
Brazil	Yes	3
Mexico	Yes	3
Kenya	Yes	4
Morocco	Yes	3
Bangladesh	Yes	2
Philippines	Yes	3

Scale of emphasis 1. strong, 2. considerable, 3. moderate, 4. negligible, 5. none

Conclusion

Although a larger sample would be needed for drawing more definitive conclusions, the survey illustrates the fact that the integration of environmental education and training, in the context of national environmental policies in the sampled countries, can play a relevant role in enabling them, as developing nations, to effectively deal with global environmental changes. Although training needs are many (see Appendix 1), such programmes are seen to have a strong potential for improving a country's ability to deal with environmental phenomena, especially those which may have strong national implications.

A special emphasis should also be made on the need for further training in the methods through which climatologic data may be disseminated in a "easy-to-understand" language, thus enabling people, mainly the non-expert public, to understand them. In addition, ways through which sets of especially relevant information might be made more widely available should be further investigated and disseminated as part of training programmes.

The provision for systematic training may provide a country with the means for understanding the reasons underlying global environmental changes and the associated problems, which may guide the search for long-term solutions.

References

Information Unit on Climate Change (IUCC) (1993) Framework Convention on Climate Change. IUCC, Geneva

IUCN / World Conservation Union (1991) Caring for the Earth. IUCN, Gland.

Leal Filho WDS (1992) Research Needs and Research Agenda. I "International Conference on Impacts of Climatic Variations and Sustainable Development in Semi-Arid Regions". Fortaleza, Ceará, Brazil. January 27 to February 2

Leal Filho WDS (ed) (1993) Environmental Education in the Commonwealth. Commonwealth of Learning, Vancouver

Magalhaes AB and Glantz MH (1992) Socioeconomic Impacts of Climatic Variations and Policy Responses in Brazil. Esquel, Brasilia

United Nations Development Programme / UNDP (1991) Human Development Report. Oxford University Press, Oxford

Unesco (1987) Report on the Intergovernmental Conference on Environmental Education, Moscow, 1987. Unesco, Paris

Appendix 1 - Training Needs in Global Environmental Changes (GECs)

1. Training is needed on how GECs affect the environment at a local, regional, and global level. In particular, attention should be paid to semi-arid regions and to threatened ecosystems, such as the tropical forests.

2. Training on the recognition of the phenomena that lead to GECs, in the framework of which the elements which create them should be investigated and ways through which their effects may be minimized, should be sought. This particular topic may include the principle of sustainable development, as a means of combining environmental conservation with economic growth.

3. Training is needed on the methods through which climatologic data may be disseminated in a "easy-to-understand" language, thus enabling people, mainly the non-expert public, to understand them. In addition, ways through which sets of especially relevant information may be made more widely available, should be sought.

4. Training is needed on how to make data on GECs available to decision-makers in specific countries, in a easy-to-understand way, thus guiding them to plan their economic and social policies, and thus stimulating sustainable economic development.

5. Training on the methods of recognition of the environmental problems associated with GECs, using data and inputs from individual and comparative research should be instituted. These may identify the extent to which problems are interrelated. The greenhouse effect, the depletion of the ozone layer, acid rain, and certain methods of energy production and fuel use should be looked at in detail and any eventual links with GEC's should be pointed out, thus guiding the adoption of proper and effective corrective measures.

6. Training aimed at catalysing the effectiveness of current international cooperation programmes on issues related to GECs is needed. This would address the deficiencies existent today in terms of information exchange, technology transfer, and transfer of resources for research, among industrialized and developing nations, with a view to improving and facilitating the access and proper exchange of valuable data.

7. Training is needed on the understanding of the phenomena related to GEC's, such as storms, inundations, cyclones, droughts, and frosts. Data on their origins, impacts, and effects should be provided during training, thus preparing trainees to identify eventual common patterns.

8. Training is needed on the measures aimed at the prevention and/or minimization of environmental damage caused by GEC's. This particular topic should be reinforced to provide suggestions of actions that should be taken at a short- and long-term level when a certain event (e.g., a rise in sea level) or series of events, occur. This is especially relevant to developing countries which often lack the means to cope with sudden changes.

NASA/USRA Cooperative University-based Earth System Science Education Program

George W Fisher
Morton K Blaustein Department of Earth and Planetary Sciences
The Johns Hopkins University
Baltimore MD 21218, USA

In an effort to strengthen the teaching of earth system science at the university level, the Universities Space Research Association (USRA) and NASA have developed a cooperative program called Earth System Science Education (ESSE) at 22 universities in the United States (Table 1). Begun in 1991 and scheduled to last through 1996, the program is intended to strengthen interdisciplinary earth science education in areas relevant to NASA's Mission to Planet Earth, to promote cooperation between NASA and university faculty, and to attract students and faculty to earth system science.

The core of the program is a pair of undergraduate courses that each participating university agreed to develop: a general survey course accessible at the freshman level, and an advanced course for senior science majors. The advanced courses were expected to involve faculty from at least two departments related to earth system science and were intended to provide a basic grounding in system concepts and models. Most universities responded by developing courses in which students from different backgrounds worked together in interdisciplinary teams to model components of the earth system. The course content tended to reflect the strengths of each university and so varied substantially from institution to institution. In return, the ESSE program provided modest financial support (for teaching assistants, a small amount of faculty release time, instructional materials, and so on) for each university, plus some additional travel funds to encourage the use of lecturers from other ESSE institutions, both to broaden the range of expertise available at each university and to promote collaboration among ESSE faculty.

In order to facilitate the program, ESSE sponsored a variety of meetings for representatives from participating universities. Steering meetings, attended by two representatives from each university, were held twice yearly to compare strategies, successes, and problems, and to share teaching materials. Given the difficulties of developing interdisciplinary programs on many campuses, these meetings proved very useful by providing a variety of program models several of which are presented in the Appendix to this volume. Two workshops devoted to exploring specific topics were also held. One was devoted to Stella[1] (a graphical

Table 1: Universities participating in the ESSE program

Universities
University of Alabama
University of Alaska - Fairbanks
University of Arizona
University of California - Los Angeles
University of California - Santa Barbara
University of Florida
University of Iowa
Johns Hopkins University
University of Minnesota
University of New Hampshire
New York University
Northwestern University
Ohio State University
Pennsylvania State University
Princeton University
Purdue University
Rice University
Rutgers University
Stanford University
Utah State University
Washington University
University of Wisconsin

differential equation solver for the analysis of dynamic systems), and the other to Geoscope (a CD-Rom-based hypermedia encyclopedia of global and regional earth science data sets developed by the Canadian Space Agency for ISY '92). Both workshops were highly successful, and facilitated the spread and development of computer-based material for teaching earth system science. These materials have proved so useful that USRA is now in the process of establishing a server to make the materials available over Internet to interested faculty at non-ESSE universities.

We have the impression that ESSE has been very worthwhile for the 22 universities participating in the program. It has clearly enabled those universities to develop undergraduate programs in earth system science education more rapidly than they would have done without the ESSE program. It has promoted the development of instructional materials and the use of

1 Stella is a trademark of High Performance Systems, Inc., 45 Lyme Road, Hanover NH 03755.

programs like Stella and Geoscope, and it has enabled those universities to develop or enhance links with the NASA community.

After participating in the ESSE program for three years, I have three observations about teaching earth system science at the university level:

- A variety of curricular and programmatic models work well as vehicles for teaching earth system science, and it would be a mistake to attempt to impose a standard earth system science curriculum on the community. Universities have different strengths and different areas of faculty interest, and each should be encouraged to develop programs which build upon its own strengths and areas of expertise.

  There is an urgent need for development of curricular materials which are responsive to the diversity of both subject matter and depth of presentation in earth system science programs. Textbooks covering the full range of earth system science topics may be less useful than short books dealing with specific topics, which may be combined by an instructor in order to meet the needs of an intended course. Some publishers are now offering to combine chapters, which they have published, to form books tailored to a specific course. This style of publication seems particularly well suited to earth system science courses. Hypertext materials available in CD-Rom or on-line via Mosaic or other Internet browsers provide an interesting alternative for those with access to the necessary technology.

- Most college professors were trained in individual disciplines, and so there is a special need for short courses in earth system topics designed for instructors, who would like to be able to teach in disciplines in which they have little background.

- There is a need to involve the solid earth science community more fully in earth system science. The geologic record provides a valuable record of the way the earth system has behaved in the past, and solid earth scientists are expert in several areas relevant to earth systems: geochemistry, rock-water interaction, sedimentation, and so on. One of the most neglected areas is that of soil development. Soils are the interface between the solid earth and the rest of the earth system, and it will be very hard to link the solid earth to the fluid envelopes and the biosphere without understanding soils better than we do now.

Global Change Education: The Evolving U.S. Federal Plan

Lynne M Carter, co-chair Education and Communications Working Group
Graduate School of Oceanography
University of Rhode Island
Narragansett, RI, USA 02882

The scientific study of global change has advanced our abilities to understand and predict environmental causes and effects, both natural and anthropogenic. The U.S. approach to the complicated science issue of global change comes through the Subcommittee on Global Change Research (SGCR), a subcommittee of the Committee on Environment and Natural Resources (CENR) representing 18 federal agencies. The members of the SGCR believed that the long term scientific study of global change required informed and educated participants. The SGCR established an Education and Communications Working Group (formerly the Education Task Group) to develop an education plan with a primary focus on the need for future scientists. However, as the questions about global change cause, effect, and policy options have evolved, so has the need for education opportunities. Preparing future scientists is referred to as a *pipeline* issue when the level of final participation is as active scientists and engineers. The first global change education program was featured in *Our Changing Planet* (1991). While identification of the need for a broad base of scientific literacy and an informed citizenry and decision-makers was briefly mentioned in the Future Program Needs section of the report, the short-term identified focus was

> To inspire and train the scientific cadre required to carry out the U.S. Global Change Research Program

It was envisioned that the Education Program would take the form of postdoctoral fellowships, graduate student fellowships, undergraduate research experience, high school summer internships, research training group awards, and summer institutes and training workshops. All of these activities would occur through the National Science Foundation (NSF) grant process and be managed by a governing board representing the SGCR global change agencies.

A number of agency activities in global change education were also underway. They included mostly graduate programs through the Department of Energy, National Aeronautics and Space Administration (NASA), Department of Interior, National Science Foundation (NSF), and National Oceanic Atmospheric Administration (NOAA).

NATO ASI Series, Vol. I 29
Global Environmental Change in Science:
Education and Training
Edited by David Waddington

The support mechanisms that were envisioned for the global change education initiatives were twofold: agency programs that were underway, and CENR initiative funding. As it turned out, only activities that were agency funded actually occurred.

All of the on-going and planned Education and Communications Working Group and agency global change education opportunities were intended to complement the science priorities of the SGCR. Those priorities were in climate and hydrologic systems, biogeochemical dynamics, ecological systems and dynamics, earth systems history, human interactions, solid earth processes, and solar influences.

The FY '92 Plan

The education plan as outlined in *Our Changing Planet: (1992)* identified the same pipeline issues and agency programs already underway and related to global change as in the *FY '91 Plan* (1991). The FY '92 program clearly identified the education planning effort as *in progress,* with a small amount of funding available for pilot projects in both the FY 1991 and 1992 years. Future education programs were expected to

> further enhance programs in the areas of research experiences, educator training, and environmental education and outreach activities at all levels.

The additional focus of environmental education and outreach at all levels was reflective of the growing interest in global change, environmental, and science education for all levels of the public.

The issue of broadening the base for global change education to include those before and following a collegiate experience leads to global change education for non-scientists but also back to the pipeline issues. Many have discussed various projections concerning the imminent national crisis in the supply and demand for scientists and engineers. Some of this shortfall is being addressed by an increasing number of foreign-born students, but that is not likely to increase enough to make up future supply needs. Although there may be enough qualified students to avoid this problem, they are not going on to college from high school (Atkinson 1990). Special focus education, aid, and grant programs for the pre-college level could encourage continued education of those specially qualified in science, math, and engineering. We cannot expect to improve the projected shortfall of scientists significantly without a continuing stream of students.

The FY '93 Plan

Global Change issues provide a relevant theme for scientific inquiry and for the policy issues that are becoming important elements in an increasingly technical society. In recognition of the relevance of global change and of a federal role in science and mathematics education in general, a broadened committee scope was developed. The Working Group long-term global change education strategy now included the need for *multidisciplinary studies, transgenerational opportunities, long-term commitments to change,* and *coordination with other relevant national and international education programs.* As a result, any global change long-term education plan needed to include opportunities for all education levels: precollege and early college, upper-level college and graduate school, and informal continuing/public education (Task Group on Education Workshop 1993).

The Education and Communications Working Group also developed a series of short-term recommendations. These short-term recommendations were meant to close the gap between what was presently being offered in global change education and the long-term strategy. The short-term recommendations for FY '93 were to: 1) focus highest priority on enhanced educational efforts for K-14/16 and on public education (since most of the present effort was for upper level science major college and graduate students); 2) utilize in-place education mechanisms in new and innovative ways (since it was unlikely that the education system could be totally reinvented); and 3) provide a central clearinghouse on global change education programs and information (since such programs are many, varied, and difficult to locate). The support mechanisms envisioned for global change education initiatives were threefold: agency global change focused initiatives; Working Group initiatives; and portions of agency environmental education and other math and science funding that was focused on global change issues (Task Group on Education Workshop 1993).

FY '94 and '95 Activities

In support of the long-term strategy and short-term recommendations developed, the Education and Communications Working Group proposed the first national global change education initiative entitled *Project Earthlink* (Our Changing Planet 1995). *Project Earthlink* will span FY '94 through FY '96. A major public kick-off will occur on the 25th Anniversary of Earth Day, April 1995. The mission of *Project Earthlink* is to: build national capacity among educators for developing global change education programs; facilitate community access to global change information; and increase scientific and technologic literacy of targeted audiences. The message is, "Our individual and collective actions change the global environment" (Project Earthlink 1994). The scientific understanding promulgated is based on IPCC and other scientific assessments (IPCC 1992).

Project Earthlink activities are aimed at six audiences: community leaders, informal educators, teachers and students, journalists, students and leaders from other countries, and the general public. Activities to reach these groups include (but are not limited to) workshops for journalists, teachers, informal educators, and community leaders; satellite video conferences; a national conference for educators, scientists, and policy makers; demonstrations of *information super-highway* access to global change data sets; and student science fair projects on global change (Project Earthlink 1994). A variety of special materials appropriate for the different audiences are being developed. For example, an educator resource guide covering natural variability, the greenhouse effect, sea level rise, ozone depletion, ecosystem response, health effects, and decision-making under uncertainty has been developed and is awaiting printing.

Other celebratory and educational activities will occur in conjunction with the 25th Anniversary of Earth Day in April 1995. An example is the year-long educational project, *Building a New World,* that will culminate in the construction of a 45-foot diameter globe by school children from the U.S., Canada, and Mexico on the Washington D.C. Mall (Project Earthlink 1994). This will also coordinate with the National Science Foundation's Science and Technology Week for 1995.

There has been a traditional education perspective to this Working Group. Since its name change to include Communications, the Working Group has redefined its vision, mission, and goals. The proposed vision is that *people of all ages will have a broad-based understanding of global change enabling informed choices leading to a sustainable environment.* The proposed missions are: *to promote understanding of global change issues leading to informed decision making, and to promote long-term formal and life-long informal learning programs in global change education.* The Working Group will continue to *oversee agency activities, support interagency projects, encourage additional agency activity in global change education, and provide federal support, mandate, and encouragement for education toward a sustainable environment* (Education and

Communication Working Group 1994). The Working Group, and correspondingly the US federal global environmental change education efforts, will continue to evolve in programs, focus, and activities, as the issues surrounding global environmental change become better understood.

References

Atkinson R (1990) Supply and Demand for Scientists and Engineers: A National Crisis in the Making, Science 248 : 427.

Education and Communications Working Group (1994) Minutes for May 17, 1994. US Global Change Research Program, US Department of Agriculture, Global Change Office, 1621 N Kent St., Rm 61 LL, Arlington, VA 22209, USA.

IPCC (1992) Climate Change. Intergovernmental Panel on Climate Change. WMO/UNEP, Cambridge University Press.

Our Changing Planet (1990). Our Changing Planet: The FY 1991 Research Plan. The U.S. Global Change Research Program, A Report by the Committee on Earth and Environmental Sciences, Global Change Research Information Offices, 1825 K. St. NW, Suite 805, Washington DC 20006, USA.

Our Changing Planet (1992): the FY 1992 Research Plan. The U.S. Global Change Research Program, A Report by the Committee on Earth and Environmental Sciences, Global Change Research Information Offices, 1825 K. St. NW, Suite 805, Washington DC 20006, USA.

Our Changing Planet (1995) Our Changing Planet: the FY 1995 U.S. Global Change Research Program, A Report by the Committee on Environment and Natural Resources Research of the National Science and Technology Council, Global Change Research Information Offices, 1825 K. St. NW, Suite 805, Washington DC 20006, USA.

Project Earthlink (1994) Private communication.

Task Force on Education Workshop (1993). Task Group on Education Workshop Report: Framework for Long-term Planning and Short-term Program Recommendations FY '93. Subcommittee on Global Change Research, US Global Change Research Program, US Department of Agriculture, Global Change Office, 1621 N Kent St., Rm 61 LL, Arlington, VA 22209, USA.

Visiting Lecture Schedule, Spring 1991

Science Issues
Overview of Global Environmental Change: The Science and Social Science Issues
William W. Kellogg, Senior Scientist (retired), NCAR

Atmospheric Chemistry: Causes and Effects
Sherwood Rowland, University of California, Irvine

The Role of the Oceans in Global Environmental Change
Berrien Moore III, University of New Hampshire

Climate and Society
What is the Role for the Press in Global Environmental Change?
Charles Alexander, Senior Editor, Time Magazine

Population and Global Environmental Change
William Clark, Kennedy School of Government, Harvard University

Energy: Its Use and Effects
Irving Mintzer, Center for Global Change, University of Maryland

Agriculture, Forestry, and Food Security
Gary Evans, U.S. Department of Agriculture

Tropical Forests
Pedro Sanchez, North Carolina State University

Potential Health Effects of Global Environmental Change
James Fouts, National Institute of Environmental Health Sciences

Policy Options and Mechanisms
Mitigation Options and Economic Consequences
Alan Miller, Center for Global Change, University of Maryland

National Legislation and Other Options
Claudine Schneider, Congresswoman (R), Rhode Island

International Treaties and Activities
Alan Hecht, Environmental Protection Agency

SECTION 2: GLOBAL CHANGE SCIENCE COURSES

Teaching teachers about global change science: Project COPE

Richard C J Somerville
Scripps Institution of Oceanograph
University of California, San Diego
La Jolla, California 92093-0224, U.S.A.

Project COPE is the name of a unique science teacher training program which took place in 1991 and 1992 at the University of California, San Diego (UCSD). The acronym 'COPE' stands for Change on Planet Earth, and the basic concept of this novel effort was to teach school teachers about the interdisciplinary science of global change. The core of the project was a series of lectures that I gave to teachers who teach science in elementary and middle-junior high schools. The grades taught by these teachers were kindergarten through eighth grade (K-8) and the ages of their students were approximately 5 through 15 years. Project COPE was sponsored by the U.S. Department of Education and administered through UCSD Extension, a part of my university which specializes in outreach activities and which offers many kinds of both conventional and non-traditional instruction for a wide variety of students.

The teachers came to the UCSD campus for one Saturday each month during 1991, except during July and August. On each of the ten Saturday mornings, I lectured to them for 3 hours, while afternoons were devoted to laboratory sessions arranged by grade level, and designed and run by science teaching experts. The laboratory sessions provided the teachers with a series of experiments, demonstrations, and other activities, which could be used in their classrooms. The lectures were intended to inform and stimulate the teachers, rather than to provide them with specific material which they could directly use with their own students. The lectures covered some of the most prominent global problems of contemporary earth. Each science, including the ozone hole, the greenhouse effect, acid precipitation, air pollution, and the loss of biodiversity. These topics are interesting and important to a broad spectrum of people today, and they are also active areas of scientific research.

The essential purpose of Project COPE was to augment the content of the curriculum and enhance teachers' instructional strategies. The idea behind the project was very simple: that earth science is alive and exciting today, and that informing teachers about current research on topics of great public interest

NATO ASI Series, Vol. I 29
Global Environmental Change in Science:
Education and Training
Edited by David Waddington

would help them convey the fascination and power of science to their students. I am convinced that the effort to increase science literacy in the general population must begin while children are very young. Science is not dull, and it need not be perceived as difficult. There is a great deal of evidence that many young children acquire a strongly negative impression about science in the early stages of their education, in part because their teachers are often inadequately equipped by their own training in science and mathematics. Project COPE was an effort to light one small candle, rather than curse that particular darkness in our education system.

How did some 60 kindergarten through eighth grade teachers come to assemble once a month on the UCSD campus to hear a university professor talk about global change? This unusual sounding project actually evolved quite naturally. UCSD Extension is a highly entrepreneurial part of the university, and it has a long history of organizing highly successful short courses for teachers. I had taught for several years in a series of summer science teacher training institutes offered through UCSD Extension. These institutes were largely inspired and organized by Paul Saltman, a biologist who is a colleague of mine on the UCSD faculty and a master teacher himself. The concept of a more elaborate science teacher training program, based on current scientific research on earth science topics of widespread general interest, appealed greatly to me. In a series of discussions in 1989, Dr. P.A. Moore, an education specialist with UCSD Extension, worked with me to plan an ambitious effort, partly modeled on these summer institutes, but stretching over a full two years and focussing on global change science. She found a possible source of financial support in the U.S. Department of Education, and together with Professor Saltman, we submitted our proposal through UCSD Extension in February 1990. It was approved in June of that year.

In the next few hectic months, we planned all the details of the program and advertised it widely in the school systems of San Diego and Imperial counties, which are two large counties in southern California near the Mexican border. Many teachers responded, and it was difficult to narrow the field to about 60. This limit on the number of participants was due mainly to the amount of money in the grant awarded by the Department of Education. Those funds were used mainly to buy materials and to pay the staff who administered the program and ran the laboratory sessions.

In teaching in the COPE Project, I quickly developed a deep respect for the teachers who were my students. They were all accomplished professionals in their own field, and they brought a rich spectrum of real world classroom experience to our program. Their energy and dedication were impressive. The great majority of them, however, were not especially well informed about many aspects of science, despite their enthusiasm for it. Nearly all of them also shared a general distaste for mathematics. In some of them, a real fear of mathematics was sadly apparent. I must also say that, taken as a group, these teachers were well aware that their own background in science and mathematics was considerably weaker than they would have preferred. This was not the fault of the teachers, After all, they were themselves products of the American system of higher education. It is a sad fact that this system produces university graduates who, more often than not, are literally unable to give a more-or-less correct answer to the question, "Why is summer warmer than winter?" As F. Sherwood Rowland has noted (Rowland, 1993), the preferred wrong answer is that the earth is closer to the sun in summer. I have given this basic science literacy question as a litmus test to UCSD undergraduate students, and I tried it on the Project COPE participants as well. The result is always the same: most people just do not know. If anyone doubts the need for improved science education, that fact should help make the case.

We met on campus on the second Saturday of every month that year, except for July and August. During the second year of the project, there were no formal lectures. There were more field trips, however, and emphasis was placed on the role of the participating teachers in staff development within their schools and districts.

Project COPE was intended to enhance the earth science knowledge base and teaching strategies of K-8 educators and to assist teachers in conducting in-service programs on contemporary earth science issues in their schools and school districts. The focus of Project COPE was to trace the scientific investigation of patterns of change in the global environment.

To increase the likelihood that school teachers would derive practical benefit from project COPE, the instructional content and laboratory activities of the project were based on the Science Framework for California Public Schools, also known as the California Science Framework, or simply the Framework. This document, adopted by the state in 1989, helps to unify systematically many aspects of teaching science. The Framework "calls for a new dynamic in science learning." Project COPE was consciously and explicitly designed to be consistent with this framework.

The Framework is detailed in a 220 page book bristling with pedagogical detail of interest primarily to professional educators. One feature strongly emphasized in the Framework seemed particularly appropriate as a guide to the Project COPE lectures. This feature was called "The Major Themes of Science."

In the words of the Framework, "Themes are the big ideas of science, larger than facts and concepts: they link the theoretical structures of the various scientific disciplines. Themes are a way to integrate the overarching concepts of science into a curriculum, much as theories encompass and connect the basic data and evidence of science."

The Framework recognizes six major' themes:

- energy
- evolution
- patterns of change
- scale and structure
- stability
- systems and interaction

Quoting from the Framework again, "The use of conceptual themes will not by itself solve all the problems of science education. However. the incorporation of themes in thoughtful and judicious ways should improve the integration of facts and ideas, the interrelationships of theories and disciplines, and the quality of instructional materials." These themes proved to be very helpful in selecting and organizing the material for the lectures, and in allowing the teachers to relate global change topics to the other subjects they taught.

Teachers who participated in Project COPE received five quarter units of UCSD credit, a stipend to help cover their expenses, and a considerable amount of documentation. We relied especially on *The Next One Hundred Years* (Weiner 1990), which became the unofficial textbook for Project COPE, and on two special issues of *Scientific American* - "Managing Planet Earth" (September 1989) and "Energy for Planet Earth" (September 1990). Jonathan Weiner's book is an excellent popular account of many aspects of global change science, and it is especially good at conveying a picture of research as an exciting activity carried on by people who are often interesting human beings! Weiner paints an engaging picture of many of the scientists who have pioneered this field, such as Charles David Keeling, whose careful measurements of atmospheric carbon dioxide concentration have rightfully become world famous as a foundation for the realization that human activities can affect our planet on a global scale.

The topics covered in the COPE lectures were varied. We covered two in considerable detail: the depletion of stratospheric ozone (dramatized by the ozone hole over Antarctica), and the potential for global climate change due to human activities, particularly by an increase in the greenhouse effect. Through these two topics, we explored many aspects of the climate system that are critical to determining the state of the environment of our planet. We then moved on to briefly cover several other topics which included acid precipitation, the global aspects of air pollution, and, finally, threats to biodiversity.

The lectures were taped, and the transcribed tapes served as the raw material for a book which I wrote after the program ended. This book *Change on Planet Earth*, was printed in 1993 at UCSD using desktop publishing technology and was distributed to the participants in the program. The book is simply a reworking of about 30 hours of my lectures. It is in no sense a complete account of Project COPE. It contains nothing of the laboratory sessions or the field activities, and it omits entirely any account of the work the participants did in their own classrooms or with their colleagues in their schools. In that respect, it leaves out evidence for how well Project COPE succeeded in its primary goal of helping teachers to improve science teaching.

Even as a record of the lectures, this book is incomplete. The transcripts of the tape recordings managed to capture most, but not all, of what I actually said. Then a skilled and patient editor, Phil Lauder, worked closely with me to translate the rhythms and conventions of spoken English into something approximating readable text. In this painstaking job, we tried to improve organization and clarity while retaining something of the informality of the lectures.

However, some special aspects of the lectures could not be fitted into this format. One was the large collection of slides, overhead transparencies, and handouts that were part of every lecture. To help compensate for this omission, I added an extensive annotated bibliography to the book to help guide readers into the rich literature of this subject.

I also decided to use the Project COPE experience as the basis for a new book on global change science aimed at the general reader. For this purpose, over a hundred copies of *Change on Planet Earth* were distributed to scientists and educators for review, many of whom sent back extensive comments and suggestions. These critiques are being incorporated in the new book. The title is taken from a poem by Elizabeth Bishop which is especially relevant to global change, *The Forgiving Air*.

Personally, I greatly enjoyed the experience of Project COPE. For me, the most rewarding aspect of the program was the interaction with the COPE participants. Self-selected and highly motivated, they enthusiastically asked questions, challenged one another and me, held spirited discussions before and after our sessions (and sometimes during them), and brought back valuable lessons from their classroom experiences. Their energy and commitment to their profession were impressive. Project COPE was an immensely rewarding experience to me, because the participants made it so.

It was a somewhat difficult experience, too, because at the outset of the course the participants established two rules for me to follow. I was told to avoid scrupulously the technical jargon that is the natural dialect of a working scientist, and I was warned not to employ any mathematics whatsoever. Forcing a scientist to lecture under these restrictions is a little like asking a cat to bark. I managed to break both of these rules. Technical terms arrived in our classroom uninvited and were introduced politely, one at a time, with proper definitions, and the participants soon became adept in the jargon in spite of their initial fear of it. Then, somehow, after a lot of numbers, one little algebraic question managed to sneak into the lectures, and I proudly included it in *Change on Planet Earth*.

The equation, I=PAT, is due to Paul Ehrlich and his collaborators, who use it as a simple model to illustrate the fact that the impact (I) of human beings on the environment is related to the size of the population (P), the affluence (A) of the people, and the technology (T) which they use. Obviously, it is a highly idealized description of reality, but it is pedagogically useful in many ways. Nevertheless, a substantial number of the COPE participants found this equation an obstacle rather than an aid to understanding. This aversion to even the simplest mathematics shocked me, and I still wonder how well a person can appreciate, say, a computer simulation of the circulation of the global atmosphere and ocean, if that same person is truly perplexed by I=PAT. Nevertheless, I may leave the equation out of *The Forgiving Air*. In his best-seller, *A Brief History of Time*, Stephen Hawking remarks that he had been advised that every equation he put in would halve the sales of the book, and that may well be so. Nevertheless, Hawking did dare to include one equation, $E=mc^2$, and I take that as a good omen for the future of science literacy.

References

Rowland FS (1993) President's Lecture: The Need for Scientific Communication with the Public. Science 260: 1571-1576

Somerville RCJ (1995) The Forgiving Air. University of California Press

Weiner J (1990) The Next One Hundred Years. Bantam Books, New York

What is Environmental Chemistry?

J Buffle and M Filella
Department of Inorganic, Analytical and Applied Chemistry
University of Geneva
Quai Ernest Ansermet 30
CH-1211 Geneva 4, Switzerland

> If you ask a chemist to find for you what a dynamo is, the first thing he would do is to dissolve it in hydrochloric acid *(Albert Szent-Gyorgyi, 1893-1986; biochemist, Nobel prize for Medicine 1937).*

Introduction

Not all scientific disciplines have reached the same stage in their development. Environmental chemistry today parallels the emergence in the nineteenth century of chemistry as a rational discipline and even more closely that of biochemistry (formerly called biological chemistry) at the beginning of this century. Achievement of scientific knowledge includes several steps: collection and classification of data, interpretation and generalization, and formulation of new, more rigorous, and quantitative concepts. Both chemistry and biochemistry have passed through long descriptive periods in which important discoveries led to the deduction of the properties of elements and compounds. Following this came a quantitative formulation of modern concepts which allowed predictions to be made on compound reactivity. In chemistry, this descriptive phase has probably reached its pinnacle, but in biochemistry there is still a great deal to be done. This lag is due to the greater complexity of the biochemical systems studied, particularly those of supramolecular organized structures.

Environmental chemistry deals with even more complicated systems than biochemistry, and it is a much younger discipline. Therefore, despite the development of modern sophisticated analytical techniques, it is still in its infancy, i.e., at the experimental data collection and classification stages. For instance consider sediments, soil aggregates, or humic compounds which are key environmental components, but whose rigourous chemical description is extremely difficult. This enormous difficulty does not mean that rigorous description of environmental systems is impossible. Biochemistry has succeeded in describing in detail the primary to quaternary structures of proteins, polysaccharides, or DNA, and even more complicated structures like biological

NATO ASI Series, Vol. I 29
Global Environmental Change in Science:
Education and Training
Edited by David Waddington

membranes, even though such achievements were considered to be impossible not so very long ago. Undoubtedly, environmental chemistry will succeed in describing the components and structures that play key roles in environmental processes, but to achieve this, as much time and effort as have been put into other disciplines will be required, and new sophisticated technologies will need to be developed to overcome the problems posed by the enormous complexity of environmental systems.

At present, environmental chemistry faces a peculiar and difficult situation. As mentioned above, classical chemistry (i.e., chemistry of small molecules of molecular weight < 1000 Dalton) has largely overcome its descriptive phase so that most chemists are trained to think in terms of predictive quantitative concepts, while the description of environmental systems and their supramolecular structures still needs semi-empirical approaches. It is therefore understandable that some classical chemists look with scepticism and sometimes with disdain on the work of environmental chemists. They should realize however that the relatively high predictive level of classical chemistry is limited to homogeneous systems that involve a restricted number of small molecules. This is at odds with most situations of the real world (industrial, life, and environmental processes), which include physically and/or chemically heterogeneous systems, and where supramolecular structures play an important role.

Like biochemistry, environmental chemistry should rather be considered by classical chemists as a challenge and an opportunity to make further progress by applying existing concepts and developing new ones for real systems.

The purpose of this paper is to discuss the specificity of environmental chemistry as compared to other environmental sciences and to classical chemistry.

The place of environmental chemistry in environmental sciences

Not only are environmental sciences in their infancy, but during the last decade their evolution has shown a leaning towards empirical studies. Indeed, because of the urgency to solve pollution problems, most administrations and many research foundations favour the funding of strategic, short-term, and therefore empirical studies. There is no doubt that urgent problems must be solved, but the authorities should not forget that for the long-term, basic research results in the most useful and the biggest pay-off. But for the results of long-term fundamental studies in biochemistry and microbiology, it would not now be possible to combat tuberculosis, diabetes, cancer, and many other illnesses. By favouring short-term empirical studies in the environment, thorough-minded scientists are discouraged from entering this field, thus hampering environmental science from developing into a sound science. This may have dramatic consequences for future generations which will be confronted by more serious pollution problems than ours, but which will have no rigorous scientific discipline to solve them.

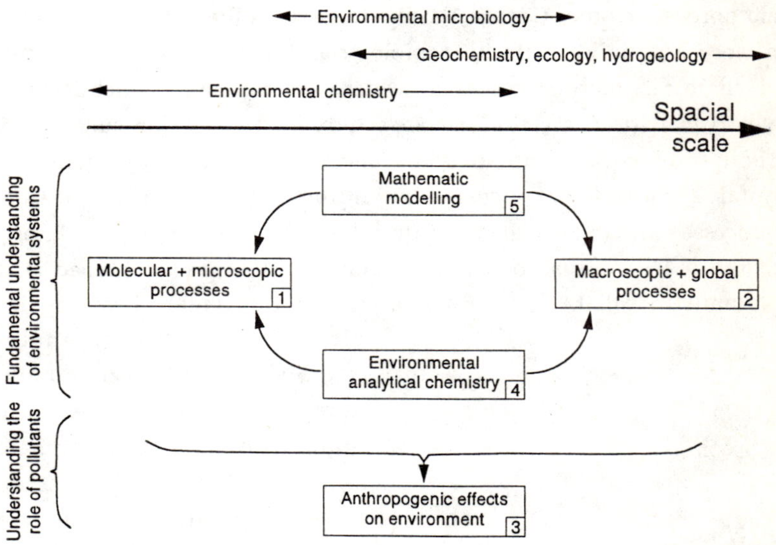

Figure 1. Different aspects covered by environmental natural sciences

Environmental chemistry should not be confused with pollutant analysis and control. On the contrary, environmental chemistry is a multifaceted discipline. A simplified but useful representation (Fig. 1) distinguishes among three aspects of all environmental natural sciences: the fundamental understanding of the functioning of unperturbed environmental systems (boxes 1 and 2), and the understanding of anthropogenic perturbations on these systems (box 3). It is essential to make the distinction between aspects 1 and 2 on the one hand and 3 on the other: anthropogenic perturbations cannot be understood and controlled without a preliminary understanding of the behaviour of unperturbed ecosystems (see below the analogy with medicine, Fig. 2). It is also useful to discriminate between environmental "microprocesses" (occurring on small scales of nm to mm) and "macroprocesses" (large scales of mm to km) corresponding to whole ecosystems. Although the border between these two scale ranges is arbitrary, the nature of the corresponding processes, the techniques used to study them, and, consequently, the corresponding conceptual approaches are very different. Teaching environmental sciences should cover all aspects of Figure 1, linking each particular discipline more closely to one or other of the aspects shown.

As far as *environmental processes* are concerned, environmental chemistry is closer to microprocess studies (Figure 1). Figure 1 shows that such studies should be done in close association with microbiology and geochemistry. In order to understand chemical microprocesses however, environmental chemistry also has to develop *analytical chemical methods* (box 4) needed for such studies. These methods are required not only by chemists, but also by many other environmental scientists since chemistry influences many geological and biological processes. Analytical chemistry has thus a key role to play in almost all aspects of Figure 1. Because of this, the general idea has developed in the scientific community that the only role of environmental chemistry is to develop chemical analysis methods for environmental quality control. Although unquestionably environmental analytical chemistry is a crucial issue for all environmental scientists, environmental chemistry should not be confined to chemical analysis. The parallel development of analytical chemistry and process studies is as vital for solving environmental problems as for combating illness in the medical domain. Figure 2 shows the obvious analogy between environmental chemistry (study of processes), monitoring, and management on the one hand, and biochemistry, medical analysis, and health control, on the other.

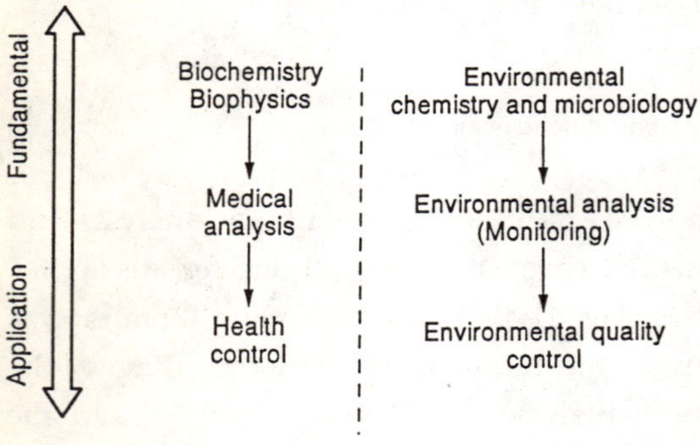

Figure 2. Analogy between environmental and medical sciences.

Specificity of environmental chemistry as a chemical discipline

The aspects related to environmental analytical chemistry will not be discussed here. Analytical chemistry - development of new analytical methods based on new physico-chemical principles - is a well-established branch of chemistry. Let us just say that, as already discussed, environmental sciences greatly need new methods applicable to diverse environmental systems.

As far as physico-chemical processes are concerned, work conducted over the last 20-30 years by Stumm, Schindler, and others has clearly shown that global geochemical cycles of elements are ultimately controlled by the reactivity and physico-chemical properties of compounds at the molecular level, and that, without a good knowledge of the behaviour of molecules of chemical compounds, no prediction will be possible at the global level. For instance, the cycling of phosphorus in lakes (with possible concomitant lake eutrophication) depends on many physico-chemical mechanisms, such as transfer of phosphate through biological membranes; oxidation of organophosphorus compounds; phosphate adsorption on iron hydroxide; nucleation of iron phosphate; oxidation of iron(II) to iron(III); calcium phosphate crystal formation, growth, and dissolution, which itself may depend on adsorption of surface active organic compounds and so on. Each of these steps is intricately related to other physico-chemical processes in a manner analogous to the metabolic cycles in biochemistry (e.g., Krebs cycle for glucide metabolism in organisms). The primary role of environmental chemistry is to elucidate the most important of these steps, in particular those which play a rate-limiting role in the overall cycle of the element of interest. In contrast to the usual systems studied in classical chemistry, the systems in environmental chemistry have the striking feature that many properties of the systems are closely related to their chemical and physical heterogeneity. The study of physico-chemical processes therefore requires using new conceptual approaches that combine classical chemistry concepts with new ones, in particular reactivity probability concepts and supramolecular structural physical models. The main differences between environmental chemistry and classical chemistry approaches are discussed below.

a. The *chemical heterogeneity* of environmental systems is a first major difficulty for classical chemists who are usually trained to think in terms of reactivity of individual molecules. The classical approach for studying the complex environmental mixtures, regardless of their complexity, consists of characterizing all the components of the mixture. Even if one casts aside the enormous efforts required to solve this problem in this manner, the solution to the questions posed are not often obtained. Indeed, this approach ignores that heterogeneity by itself is a property of the environmental systems of interest. Fulvic and humic compounds of soil and waters provide a good example of such a case. Even after "purification", fulvic and humic samples are composed of a large number of similar but not identical compounds. As an ensemble, they act as a complexing system for metal ions (and small organic molecules), with buffering capacities spanning an extremely wide concentration range of metal ion. As such, they play a key role in maintaining acceptable concentrations of vital and toxic metal ions for organisms in soils and waters. Alone, none of the individual species will impart this property. Hence, they should be thought of as a single chemical system, rather than as a mixture of individual species. This buffering action due to chemical diversity (or chemical heterogeneity) can be seen as being equivalent to the increased stability of ecological niches when the number of biological species increase (biodiversity). It has been shown that the overall properties of heterogeneous chemical systems may be rigourously related to the properties of individual molecules by introducing the concept of reaction probability. Descriptions of such chemical systems should then be based on rigourous statistical concepts, applicable not only to a large number of identical molecules (as is usually done in classical thermodynamics), but also to a system consisting of a large number of similar but not identical molecules.

b. It is now well established that the most important reactions in environmental systems occur in solid-liquid (soil, sediment, water) and gas-liquid (soil, ocean, fog droplet surface) *interfaces*. The electric field and solvent structure are very different at these interfaces (over a thickness of tens of nm) than in the bulk solution and thus may drastically modify the nature and rates of chemical reactions. A first glimpse at many chemistry curricula for chemistry students shows that surface reactions (adsorption isotherm, interface catalysis, etc.) are barely (or not at all) taught.

c. The previously mentioned interfaces are the simplest examples of *environmental microstructures*. All environmental systems are physical structures at all levels, from nm to km range. Examples of large structures are soil horizons; lake, ocean, or atmosphere stratification; and geological layers. Microstructures are those with which environmental chemistry is more concerned. Aggregates between large polysaccharide macromolecules, clays and microorganisms in soils, water, and sediments are examples of microstructures in the range nm to mm. They play a key role in the circulation of oxygen, water, and nutrients, and therefore in the maintenance of life. Sediment-water interfaces in lakes and oceans are other examples. They are stratified at the μm to mm level by successive layers of varying chemical composition, which controls the overall exchanges of chemicals between the whole of the water body and the sediment. Similarly, the gas exchanges between the ocean and the atmosphere are predominantly controlled by a diffusion layer of 30-80 μm whose chemical composition is very different from the bulk solution. In all cases, the functioning of living environmental macro and micro systems is very much independent of their structure (see d). Indeed, living systems (organisms or ecosystems) need pH, redox, or other chemical gradients which can only develop with physical heterogeneity. A good characterization of environmental microstructures is therefore a key task of environmental chemists (as characterization of cell structure is a key aspect in biochemistry). This implies knowledge of colloid chemistry, macromolecule physical chemistry, and basic biophysical concepts. This is unfortunately at odds with the classical training of chemists which is usually restricted to homogeneous solution chemistry (see the introductory quotation by Szent-Györgyi).

d. The physical structures existing for all scales of environmental systems and the corresponding chemical gradients imply the existence of *fluxes of compounds* within the system of interest (e.g., soil aggregate, sediment-water interface) or between one system and the neighbouring systems. Environmental systems (micro and macrosystems) are therefore dynamic systems. Thus, understanding environmental processes necessarily implies coupling kinetics of chemical reactions with physical transport processes, which often occur by molecular diffusion in microsystems. This is an additional lacuna in the training of most chemistry students. Not only is the coupling of chemical and diffusion equations rarely taught, but often more emphasis is laid on chemical equilibria as compared to chemical kinetics.

It is essential to realize that properties (a) to (d) above, in combination with classical chemical reactions, are the key factors which assure the *homeostasis* of environmental systems. Therefore, they also correspond to sensitive properties of environmental systems, which may be particularly affected by anthropogenic activities. These concepts should be taught in detail to all environmental scientists.

Teaching and research in environmental chemistry

Role of modelling

In recent years, the use of numerical simulation models as a means to evaluate large scale or complex natural processes has increased sharply. In some cases, the predictions generated by these models are considered as a basis for public policy decisions (e.g., global circulation models to predict the earth's climate). There is an over-emphasis on the predictive value of modelling, often neglecting the fact that all models are oversimplified compared to natural processes, many natural processes are still unknown, and, above all, that the primary value of models is heuristic. On the other hand, it is unquestionable that models have enormous importance as educational tools. They allow in particular i) classification of information, ii) checking working hypotheses with experimental data, iii) formulation of new hypotheses by eliminating unimportant factors, and iv) sensitivity analysis.

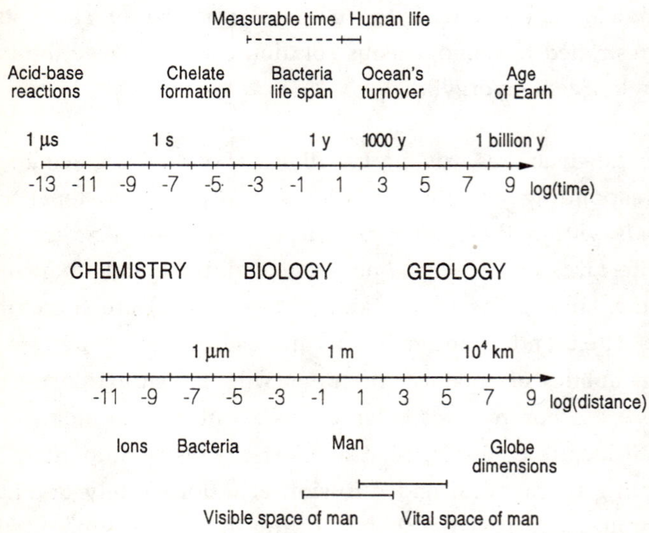

Figure 3 Time and space domains covered by environmental sciences.

These aforementioned considerations apply to all disciplines. But as a learning tool, modelling is particularly important in environmental chemistry. Probably it is the only tool available for quick evaluation of the influence of factors affecting molecular or microscopic processes in the macroscopic evolution of ecosystems. As explained later, the time and space domains of chemical processes are much farther away from those occurring on the macroscopic or global scale (Figure 3). It is thus extremely difficult, both for students and researchers, to assess the actual importance of chemical processes on ecosystem evolution, and in this respect modelling is an extremely useful tool for linking microscopic and macroscopic processes.

Experimental work

As mentioned earlier, expertise in any scientific discipline can be acquired only by working experimentally in that field for at least some period of time. This is even more important in the environmental sciences because of the complexity of environmental systems and the multidisciplinarity of the problems involved. As far as environmental chemistry is concerned, the student will have to deal directly or indirectly with three experimental aspects:

- Study of physico-chemical processes *in situ* (in the field): planning and performing well-controlled experiments in the field is much more difficult than in the laboratory, because of the large number of factors to be considered and controlled in order to allow sound interpretation of data.

- Study of physico-chemical processes in the laboratory with model compounds simulating natural systems: studies on natural systems allow one to discriminate between predominant and secondary factors affecting a particular macroscopic process. However, results of *in situ* measurements cannot provide precise values of physico-chemical parameters. These can only be determined in the laboratory by using model experimental systems that mimic natural ones.

- Development in the laboratory and application *in situ* of analytical methods specifically designed for studying environmental processes: measuring the right parameters is a key issue in order to understand natural processes. This is far from trivial in complicated natural systems.

Teaching environmental concepts

The previous section discusses the basic physico-chemical concepts which are currently missing in many chemistry curricula and are a prerequisite for understanding environmental chemical processes. Teachers in the classroom will be faced with two additional difficulties of a general type - other than those specifically related to the principles - while teaching these concepts to the students. The first difficulty is related to the time and space scales covered by the environmental sciences (Figure 3). It is not always easy for students to grasp the idea of how chemical processes occurring at the nm to mm level can affect ecosystem evolutions over centuries or even geological times. Another problem is our "sensitive" time window. Every-one understands the meaning of time flow, if the time ranges from fraction of a second to millenia. Scales such as these are met in environment macroprocesses (often biological, geological, and physical processes), where students do not have problems visualizing them. However, the time-space windows in physico-chemical processes are quite different from the latter ones, and it is difficult to help students understand their relevance in environmental chemistry (i.e., the link between chemistry and environmental processes). The situation becomes worse if one compares the "sensitive" space window of man (fraction of nm to tens of km) to the size of plankton (10^{-3} - 10^{-6} m), colloidal particles (10^{-8} - 10^{-6} m), or simple molecules (10^{-9} m). These entities differ in size by orders of magnitude, and therefore behave in a completely different manner. The importance of these differences is not obvious to the students who are new to these concepts.

A second difficulty of environmental chemistry is that many different concepts must be integrated in order to understand the behaviour of any particular environmental system. The principles of the various individual chemical reactions (acid-base, redox, complexation, precipitation, hydrolysis, etc.), that may occur in this system, are often easy to understand. It is, however, much more difficult for students to decide which processes to ignore and which ones to consider, i.e., which ones play a major role in the evolution of the environmental system of interest, in particular when non-chemical processes have also to be taken into account. In most chemistry courses, students learn to understand single processes in depth, but they are not trained to think in terms of combining many processes together. At the general teaching level this may be one of the most important impacts of environmental chemistry on chemistry students. Even if they do not pursue their studies in the environmental sciences, learning how to combine quantitatively many different factors affecting a given process will be very fruitful for any real world application.

Conclusion

The learning of a science involves more than the acquisition of new knowledge by a blank mind. Instead it involves a substantial restructuring of pre-existing knowledge, in which new knowledge must compete with students' previous knowledge and familiar ways of thinking. This is particularly true for chemistry students learning environmental chemistry and environmental sciences.

It should also be stated clearly to students that any curriculum of one to three years in environmental chemistry or environmental sciences is inadequate to make them specialists of environmental processes. The basic requirement for a medical degree is five years, and specialisation in a particular field of medicine is required: that is, additional years must be spent before one can claim to be a specialist. Environmental systems are certainly at least as complicated as the human body. There is therefore no hope of producing environmental scientists with expertise within 1-3 years. The only thing we can do is to give students the basic concepts necessary to learn so that they can progress further by themselves.

CHEMISTRY, RESOURCES AND THE ENVIRONMENT

A course for chemists interested in environmental issues including global change phenomena

C Anastasi
Department of Chemistry
University of York
Heslington
York YO1 5DD, UK

Introduction

The interdisciplinary nature of Global Change Science brings problems and challenges to those developing courses in this area. There is a strong feeling that students should be exposed to the interaction between disciplines as much as possible so that a global perspective is gained. As noted elsewhere in this book, the absence of a department in an institution dedicated to Global Change Science usually means the relevant expertise required for such a course is in institutions having departments whose focus and application of resources lie in different directions. A way forward to course development is the "staple gun" approach with its inherent strengths and weaknesses, examples of which are presented elsewhere in this book, (Schneider 1994). In many ways it is a revolutionary approach to the problem of delivering Global Change courses.

An alternative, more modest way forward for institutions that wish to educate students in Global Change issues, but whose resources are limited, is the subject of this article. It essentially champions an evolutionary approach to teaching the subject, and builds on the knowledge of the two most important groups of people: the students and the academic staff. For the students, the material presented builds on a firm and expanding knowledge of their chosen major discipline: chemistry, physics, biology and others. For the academics, Global Change issues are discussed within the context of their own expertise. The common ground for both students and academics lies in the major discipline.

The course in the Department of Chemistry at the University of York is an example of this model. In describing the course in detail, the strengths and weaknesses of the approach will be highlighted. It is a course that has been run successfully for about fifteen years and has been described as an environment course. What will become obvious is that it covers many of the elements needed to educate students in Global Change issues.

NATO ASI Series, Vol. I 29
Global Environmental Change in Science:
Education and Training
Edited by David Waddington

Key aims of the course and departmental strategy

There are three key aims of the course:

- To illustrate the challenges facing society as a result of the increasing demand for natural resources
- To show the essential role of chemistry in meeting these challenges
- To highlight the impacts of human activities on the global environment

Additionally, the course material lends itself readily to the development of other, transferable skills including communication (both written and oral), team work, problem solving and development of analytical skills beyond those found in the major disciplines.

The strategy adopted by the Department was to harness the in-house expertise available and to create a course that is concerned with *illustrating* environmental principles rather than one that seeks to cover *all aspects* of the subject. This immediately removes the problems associated with inter-departmental barriers, both technical and behavioural, that have been highlighted elsewhere in this book. Also, the resources needed are considerably reduced and the integration of material is much easier. However, the course presented is invariably less broad than may be ideal, and the student body the course services has a narrower background than might be desired.

Course Framework which Incorporates Environmental Option

The Department of Chemistry at York provides a three year degree dedicated to students wishing to specialise in chemistry and related subjects. The environmental option is taught side by side with the larger chemistry course. The objectives of the former are:

- The application of 'core' chemistry in an environmental context
- The development of environmental ideas within a secure and continuing chemistry framework
- An increasing emphasis on environmental content as the course progresses

The table below summarises the course structure. All students take the same 'core' chemistry in the first year; this part of the course provides students from different schools and curricula with the same knowledge base and prepares them for more advanced material in the second and third years.

Part I	Year 1	All students take the same (core) chemistry course.	
Part II	Year 2	Two-thirds of the course (the core) is the same for everyone.	One-third consists of two optional units.
	Year 3	One-third of the course is core chemistry.	Two-thirds consists of four optional units, a literature review and research project.

In the second year, two-thirds of the course is the same for everyone and concentrates on core chemistry and one-third of the course is made up of optional units chosen by students to suit their interests. In the third and final year of the degree programme, only one third of the course is core chemistry common to all students, while two thirds of the year is spent on optional units. During their last year students also carry out Literature Reviews and Research Projects, which are usually related to the areas covered in their optional units.

Students find the course attractive because essentially they are able to create one-third of their course through their choice of a portfolio of optional units. Some students wish to concentrate on mainstream chemistry, and so choose the appropriate optional units that allow them do this. Other students wish to study environment, industry or life science issues, and so choose the optional units that form the respective courses: *Chemistry Resources and the Environment* (described in detail below): *Chemistry, Management and Industry:* and *Chemistry, Life Sciences and Pharmaceuticals.* There is also an opportunity for students to take a selection of units from each of the course options. In this way it is possible to raise the awareness in environmental issues of a large fraction of the undergraduate school..

Environment for All

As indicated above, the unit options are available to students in their second and third year. However, there is an attempt to increase student awareness about environmental issues through workshop activities in both the first and second years. The activities are diverse in nature and are aimed at developing different skills.

An example here is the annual *Happening*. This is a workshop in three parts which involves all of the first year school. Small groups of 3 or 4 students are formed, and each group then formulates responses to specific technical questions. The first part of the workshop aims to educate students about the industrial production of chlorofluorocarbons, with particular emphasis on why these compounds were developed in the first place, and highlighting their strengths and weaknesses. In the second part of the workshop the fate of chlorofluorocarbons in the atmosphere is discussed so that students are then able to understand the impact of these compounds on stratospheric ozone and global climate change. The students are then sufficiently well informed to be able to contribute to the third part of the workshop, a debate in which two student teams present the case for and against a motion such as: *"This house believes that all man-made halogen containing organic compounds should be banned".*

This activity allows students the freedom to learn from their tutors and their peers within relatively secure technical boundaries, and in a generally informal environment. The debate provides the vehicle for using the knowledge gained in a decision making way. It also allows students, probably for the first time, an opportunity to explore the interaction between the certainties of science, and the uncertainties of emotion often engendered by environmental issues. In a broader context the workshop is an excellent vehicle for students to discuss the fundamental question of whether technology is able to provide solutions or 'fixes' to environmental issues, an aspect that is of increasing importance in Global Change Science.

The Environmental Option: 'Chemistry, Resources and the Environment

As indicated earlier, this option begins with two units in the second year, followed by four more units in the third year. The four units that form the basic environmental option along with the main aspects covered in each unit are given below:

- *Earth Sciences:* Geosphere; Rocks; Hydrosphere; Biogeochemistry
- *Energy Studies:* Energy in the environment; Fossil fuels; Electricity generation; Renewable energy sources
- *Land and Water:* Water resources; Crop production; Fuels from biomass
- *Chemistry in the Atmosphere:* Chemistry in the atmosphere; Measurement and control of air pollutants; Atmospheric composition and climate

It is clear from this description that many of the fundamental components of Global Change Science, as defined elsewhere in this book, are covered in these units, and in an integrated manner.

Two other units have a significant content relevant to those taking the environment option and complement the material given above:

- *Raw Materials, Processes and Products:* Resource utilisation; Carbon based minerals; Mineral economics
- *Managerial Economics:* Market economics, Project appraisal and risk analysis; Location of industry and the environment

In these last two units the students must rely and expand on their knowledge of chemistry as they are exposed to issues that are largely the concern of industry, yet also impact on the environment. This allows students to gain a more complete insight into the driving forces associated with the consumption of resources and their impact in global change issues.

Students with a keen interest in environmental issues will tend to take all of the optional units. Other students may take a smaller selection, and although these students have a less broad perspective of issues and lose a little in terms of the integration, they at least have been exposed to some aspects of the subject.

A breakdown of one of the units, *Chemistry in the Atmosphere,* is shown overleaf a nd serves to illustrate the breadth and depth of the material involved. The description also gives an indication of the kinds of activities the student is exposed to that enhance other skills, both analytical and behavioural.

The lectures are approximately 1 hour long. Each tutorial involves 9 hours work of which 8 hours are for completion of set work and private study, and 1 hour is spent in small group discussion with a tutor. The workshops have a time allocation of 16 hours, again spent in small groups. An example of the kinds of workshop activities is given elsewhere in this book (*The Truth About Catalytic Convertors.* A typical unit then, has at least 70 hours of formal study associated with it.

CHEMISTRY IN THE ATMOSPHERE UNIT

The chemistry of our atmosphere is an important environmental issue that has received considerable attention in recent years. In this unit the interplay between the structure of the atmosphere and the chemistry is explored, and the role of laboratory studies, ambient measurements, pollutant fluxes and modelling are discussed. Control of air pollution through legislation, and the costs involved, are also important issues addressed in this unit. The Greenhouse Effect, Acid Rain and Photochemical Air Pollution receive individual attention and are used to investigate possible approaches to air pollution in the future.

MAJOR AREAS COVERED:

Chemistry in the Atmosphere:

- Structure in the atmosphere
- Stratospheric and tropospheric chemistry
- Laboratory measurements of gas phase processes
- Liquid and solid phase chemistry in the atmosphere

Measurement and Control of Air Pollutants:

- Inventories of air pollutants
- Measurements in the atmosphere
- Modelling of air pollutants
- Pollution abatement strategies
- Legislation and policy
- Economics of air pollution

Atmospheric Composition and Climate:

- Radiation, atmospheric composition and temperature
- Climate change
- Planetary atmospheres

LECTURES: 27 over three 9 week terms (this includes some lectures on specialist subject matter given by experts from government research institutes)

TUTORIALS: 1 per term

PRACTICAL WORK: A selection of workshop activities dealing with paper comprehension, problem solving and data analysis

SITE VISIT: Visit to an urban air pollution monitoring site

As indicated earlier, an important feature of this general approach to Global Science Education from the students point of view is that they are able to use aspects of their core chemistry to support and build on in the environmental option. In the unit *Chemistry in the Atmosphere* summarised above, links to the 'core' chemistry include:

- Gas phase chemical kinetics including reaction rates, thermochemistry and mechanistic chemistry
- Liquid and heterogeneous phase chemistry
- Spectroscopy and analytical methods

There are obvious links in this unit to global change issues and examples include:

- Stratospheric ozone
- The emission, chemical transformation and measurement of trace greenhouse gases in the atmosphere
- Trace gas contribution to radiation balance in the atmosphere

The assessment of the units is in two parts. The first contribution is a continuous assessment based on the various workshop activities, including *The Truth About Catalytic Convertors. The* major assessment is through the a formal examination of the unit at the end of the course. Examples of the kinds of questions set in the climate change area for students taking the *Chemistry in the Atmosphere* unit are given below.

SAMPLE EXAMINATION QUESTION FOR THE "CHEMISTRY IN THE ATMOSPHERE" UNIT

The information summarised below is often presented in discussions of the Greenhouse Effect:

Trace Gas	*Lifetime/years*	*Global Warming Potential (GWP) relative to CO_2 (per molecule basis)*
CO_2	uncertain	1
CH_4	10	21
N_2O	150	206
CCl_3F(CFC-11)	60	12400
CCl_2F_2(CFC-12)	130	15800
$CHClF_2$(HCFC-22)	15	10700
C_2HF_5(HFC-125)	28	13400

(a) Explain why the lifetime of CO_2 is uncertain, and why calculating GWPs of trace gases relative to this species is unsatisfactory. (5 marks)

(b) For each of the gases in this table, with the exception of CO_2, write down the key reactions that determine its lifetime in the atmosphere. (7 marks)

(c) Explain the differences in GWPs shown in the table above. (3 marks)

(d) Explain why the GWP of atmospheric O_3 is difficult to include in such comparisons. (5 marks)

SAMPLE EXAMINATION QUESTION FOR THE "CHEMISTRY IN THE ATMOSPHERE" UNIT

Methane is a key component of the natural and polluted atmosphere. Its concentration in the troposphere is steadily rising, as shown in the figure below.

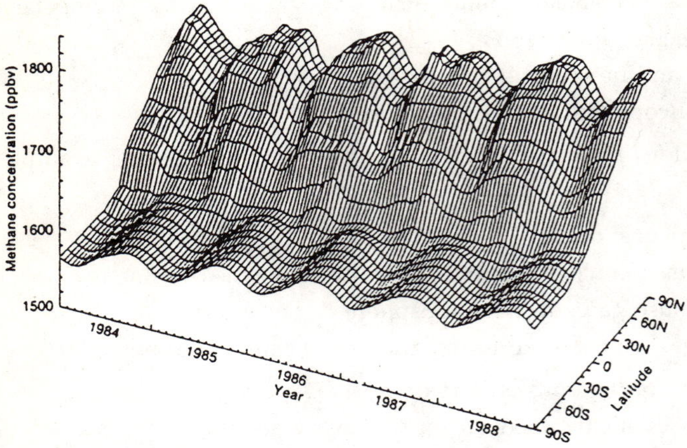

The global distribution, seasonality, and trend of methane in the troposphere

(a) State the significant sources of methane, indicating the approximate contribution each makes to the total concentration in the troposphere. (6 marks)

(b) Explain the latitudinal and seasonal variations depicted in the figure above. (6 marks)

(c) Describe the fate of methane in terms of the reactions it undergoes in (i) the troposphere and (ii) the stratosphere. (4 marks)

(d) Describe briefly the importance of methane to the greenhouse effect. (4 marks)

The examination paper is two hours long and students will be expected to attempt three such questions out of four. Each unit contributes 6% to the overall degree mark, so that in total, the six optional units that make up the environment course contribute 36% towards the final degree. Also, the literature review and research project described below contribute a further 12% to the degree. This is of considerable benefit to students since they are being assessed in a variety of ways on issues and content that they have largely chosen.

Literature reviews and research projects

Literature Reviews and Research Projects are final year undergraduate activities and are attempts to take the student into the more speculative areas of the subject. The objective of the Literature Review article is to collect into one document in a coherent way, information which has been obtained from a variety of published sources. The time allocation for the study is approximately 35 hours and the final report does not normally exceed 2500 words. Examples in the climate change area demonstrate the breadth possible:

- Absorption of infrared radiation by trace gases in the atmosphere
- Aircraft emissions and their contribution to atmospheric chemistry
- Aerosols in the atmosphere
- Feedback mechanisms in climate change

The Research Project is intended to provide the student with an opportunity to learn how to plan, carry out and report a piece of original work, and thus to develop further analytical skills. The time allocation for the study in this case is approximately 100 hours and the final report does not exceed 4500 words. Once again examples in the climate change area include:

- Free troposphere chemistry
- Absorption bands and overlap effects in infrared radiation modelling of the greenhouse effect
- Latitudinal distribution of methane and hydroxyl radicals
- Future population growth, fossil fuel usage, and the greenhouse effect

The project description for the *Free Troposphere Chemistry* research project is given next to illustrate the range of activities involved. It is common practice for the student to be immersed into an established research group with all the hidden benefits that entails.

FREE TROPOSPHERE CHEMISTRY

An undergraduate research project in the climate change area

PROJECT AIMS:

- To develop and validate a model that describes the chemistry of trace gases in the troposphere
- To use the model to assess the impact of increasing concentrations of greenhouse gases on the production of tropospheric ozone and the cleansing potential of the hydroxyl radical

STUDENT ACTIVITIES INCLUDE:

- Establish extent of knowledge in this area
- Develop chemical models for the tropospheric degradation of methane, hydrochlorofluorocarbons, and carbon monoxide for incorporating in computer models
- Establish fluxes of key pollutants
- Conduct a suitable variational analysis using computer based models
- Write a formal report on the findings of the research

In summary, the Literature Review and Research Project activities allow the student to develop research skills, to demonstrate initiative, and to take responsibility for the programme of work and its implementation. The actual review and project report also allow the student to demonstrate writing skills with particular emphasis on content, delivery, and balance. A bonus is that the student's research work can lead to a full research project, which the students themselves can develop in postgraduate programmes.

Future Development of Courses for Global Change Science

There is little doubt that the course described above has been, and continues to be successful. The Department of Chemistry at York takes approximately 115 undergraduate students each year. Approximately one quarter will take all of the environmental units and, as indicated earlier, many more students will take one or more of the units summarised above.

The course also continues to evolve. This is essential not only because our knowledge and understanding of environmental issues is increasing, but also because the character of the student intake to undergraduate degrees is changing. It is also important to accommodate the changing interests of the academics involved in the course and the introduction of new academics. These activities help to keep the course up-to-date and relevant.

The model described in this article is a relatively simple one which can be adopted by departments within which a section of the academic staff have interests in, and a commitment to, Global Change Science. The introduction of such a course can broaden the appeal of the department significantly and lead to increased student numbers. It can also evolve to match the changing needs of the subject, the students, and the academic staff.

Acknowledgements

The author wishes to acknowledge John Garratt who had the original idea for the Chemistry, Resources and Environment course at York and did much to promote its development. The author also wishes to acknowledge members of the Department of Chemistry who have contributed to the successful delivery of the course over the years including Peter Hanson, Barry Thomas, John Vernon, David Goodall, Tom Halstead and others.

References

Schneider S H (1994). This book

Teaching Environmental Sciences to Undergraduate and Postgraduate Chemistry Students

M Filomena Camoes and Leonel A T Neves
CECUL, Department of Chemistry
University of Lisbon
1294 Lisbon Codex
Portugal

Introduction

In the Department of Chemistry at the University of Lisbon (FCUL) a one semester course on "Environmental Chemistry" was introduced ten years ago for undergraduate students studying for the Educational Chemistry Diploma. The course is taken in the 6th semester, and it is an opportunity to strengthen the scientific background of students, who until then have only taken one semester of analytical chemistry and who will soon be teaching in and working on areas related to a multidisciplinary subject.

A strong component of analytical chemistry is emphasized in the environmental chemistry course, since one of the main purposes is knowing the composition of real systems.

The adopted methodology combines regular lectures, showing the relevance of a theoretical approach, with laboratory sessions and study visits, and stresses the importance of good quality data upon which wise decisions can be made. Therefore, assessment is based on performance in the laboratory, the quality of reports and written tests, and a monograph that is elaborated on by oral presentation and discussion.

Ten years ago environmental issues were becoming more and more important. It was obvious that this message should be quickly and effectively passed on to as many people as possible, so it was thought that investment should be made in the education of young children. Accordingly, the Department of Chemistry decided to promote as top priority the improvement of the scientific and pedagogic standards of the undergraduates who would soon become school teachers.

The mass media are constantly reporting on environmental disasters and people are realizing that things are not what they used to be. Water appears to be

NATO ASI Series, Vol. I 29
Global Environmental Change in Science: Education and Training
Edited by David Waddington

polluted, dirt brings trouble to the nose and eyes, the rivers and lakes are dying, the air is smelly and harmful to the lungs, and there are many allergies.

Inevitably chemistry comes up as 'the bad guy', responsible for these problems. The school teacher must have the knowledge and the skill to transmit more accurate information than that obtained from the media, not telling 'stories' about events, but helping students to determine some truth from the facts, teaching how and why things are happening, how to evaluate, how to prevent pollution, and how to recover resources. Besides teaching children, it is of the utmost importance that the role played by the teacher is extended to help people in general to understand environmental problems. The community has to be educated in order to contribute to improving the quality of life.

Chemistry is everywhere in natural phenomena and in anthropogenic ones.

Nature itself is like it is in a big beaker: matter exists in a limited amount, and happening all the time are changes that lead towards a complex physicobiogeochemical chain of events. When equilibria are established in nature, the coexistence of the constituents of the different systems work together with harmony and a special kind of beauty, but if significant perturbations are introduced, pollution occurs.

The bigger the concentration of contaminants, the higher the rates of the reactions. Flying over a big, brightly lit city such as London at night, one has a good example of what might be involved in the management of big centres in terms of water supplies, sewage treatment, and air quality control.

What may have started as a healthy attitude - one family, one house, one farm - may soon result in an uncontrolled population with needs, requirements, and deleterious outputs. Neighbours react to a smoky fire, to a contaminated water flow, to a pit, to a barking dog. Extrapolating, we have what now exists: big cities, industrial complexes, and intensive agriculture, all of which consume and produce chemicals in the big beaker.

Intensive agriculture usually violates the cycles of nature, and the industrial complexes can be responsible for the destruction of plant and animal life in extensive areas. Both of these are important parameters of the increasing pollution of the earth.

It is urgent to help people to develop a state of mind, in which the selection of materials, consumables, and ways of disposing of residues are carefully considered.

Industry must use technologies that pollute less and contribute part of its profits to the development of cleaner technologies. For the preservation of humankind and quality of life, all human beings and human institutions must be good examples of civic and professional behaviour.

Reduction, Reutilization, Recycling

It is always necessary to remember that the producers - those working in industry as well as in agriculture - have to be assisted by experts capable of advising on the optimization of processes, while the consumers must be reassured, not only of the quality of goods, but also about the processes used and the environmental consequences of manufacturing.

Often it is too late when one finds out about the consequences. It took 20 years to realize that inhalation of unseen asbestos fibres produced lung cancer. Ozone depletion was only detected after many years of use of CFC's. How many years, if at all, will it take to overcome these and other problems? In many cases, the intensive remedy, by removing and cleaning, is possible and effective, although expensive. In some instances, the attempted solution of a problem brings another, sometimes worse problem.

Education and Training

To summarize what has been said: education and training are required at various levels in order to contribute to the same big issue.

Laboratory and field technicians play a very important role, since collection of samples, performance of measurements, and keeping equipment in good working condition are essential for good quality data upon which scientists can rely. Only education at an intermediate level that is followed by appropriate training will produce competent professionals needed for this work.

Intensive courses on global environmental change are recommended for people who, because of some sudden increase in professional demands, need extra training.

The scientist, who is expected to see further, needs higher education and a diversified curriculum. Degrees in physics, biology, geology, or chemistry, with a strong component of mathematics and modelling, appear to be useful in obtaining the recommended background.

Postgraduate studies in either field must ensure the production of experts, but experts capable of leading teams.

APPENDIX

Environmental Chemistry Course : 6th Semester
Diploma : Education in Chemistry
(3 hours of lectures and 3 hours of lab work per week)

Lectures

Introduction:
- Chemistry of the environment
- Distribution of the elements
- Natural and anthropogenic pollution
- Urban centres and industrial areas
- Quality indicators
- Biological indicators
- Biogeochemical cycles
- Energy, thermodynamics, and kinetics
- Residence time
- International organisations : standards and recommendations
- Technical and scientific developments : measuring and modelling

Evolution of the Earth:
- Origin and development of life
- Major and minor components
- Speciation
- Essential and toxic elements : accumulation factor and food chains

The Atmosphere and Global Climate Change:
- The Atmosphere : composition and circulation
- Air pollution
- Aerosols
- Atmospheric photochemistry
- Fossil fuels and the Greenhouse effect
- Acid rain

Aquatic Systems and the Environment:
- Water quality and management
- Waste water treatment
- Aquaculture
- Eutrophication
- Chemistry of some elements
- Historical introduction to oceanography

The Earth's Crust:
- Soils
- Agriculture and food supply
- Mineral resources - metals and nonmetals
- Sediments
- Waste disposal

Selected Topics:
- Detergents
- Pesticides
- Polymers and plastics
- Nuclear chemistry
- Biodegradation

Laboratory:

Water analysis bulletin (organoleptic factors: phyisicochemical parameters: toxic and microbiological parameters)
DO ; BOD ; COD
Salinity and chlorinity
Sulphates in sea water
Hardness of water
Nitrates and nitrites in cold meats
Pesticides
Lead in human hair
NOx
Ozone

Study Visits:

Not more than 3 out of the following:
- Papermill
- Water treatment plant
- National Meteorological Observatory
- Oceanographic ship
- Nuclear reactor
- National Research Institute for Fisheries

The Science and Social Issues related to Global Environmental Change: an Undergraduate Honors Colloquium

Lynne M Carter
Graduate School of Oceanography
University of Rhode Island
Narragansett
Rhode Island 02882, U.S.A.

Introduction

The constant and sometimes dramatic changes in the climate of the earth have been recognized from time immemorial. What is new is the recognition that humankind has generated physical and chemical changes in the environment that can affect the climate of the natural earth system and the economy of the human earth system. Affluent societies emit carbon dioxide and chlorofluorocarbons as by-products of an increasing standard of living. Developing nations add to that burden yearly by burning forests the size of Tennessee. Population growth rates of some developing nations will push societies from living on this earth to merely surviving on it if they are sustained. We have put in motion an environmental experiment whose consequences we are only beginning to contemplate.

Assessing and understanding the causes and potential consequences of this environmental experiment are issues that need to be addressed by the physical sciences, the social sciences (political, social, human behavioral perspectives), and by innovative joint studies which include physical and social scientists. There are new physical science planning and coordinating groups, as well as major new scientific programs for understanding the causes and consequences of global environmental change. However, the complex science issues of global change, that the next generation must understand, are inexorably linked to social science issues: energy use and policies, by-products of technological development, population growth, and environmental impacts. The social aspects and concerns are far more difficult to define than the physical factors. The question "What actions can humankind implement to prevent, mitigate, or live with the scientifically projected environmental effects of global warming?" is much farther away from any real answers than, "What are the major causes of methane or carbon dioxide increases in the atmosphere?"

NATO ASI Series, Vol. I 29
Global Environmental Change in Science:
Education and Training
Edited by David Waddington

Universities and other educational facilities are being called upon to define the potential human consequences and to address the response strategies of global environmental change. The traditional university role includes at least two functions: education and research. The educational role encompasses both student training and professional development. This short paper will not address the function of research or professional development in any detail. We must, however, remain critically aware that professional development is important for those who will not become professional scientists, but professional citizens whose daily choices and participation in environmental policy will direct society's course. As we can see, that course has important ramifications for environmental sustainability.

We have spent our educational time wisely only insofar as preparing individuals to deal with specific disciplinary realms and questions, but we are only now recognizing the need for educational programs that go beyond classical single discipline courses in bridging the gap between the physical and social aspects of this potentially all-encompassing issue. It is crucial that we now include both the single disciplinary educational opportunities and the interdisciplinary educational opportunities in higher education. One of the challenges to humankind today is to prepare the members of the next generation, by stretching their limits to be more capable of effectively dealing with the complex science/social issues that make up the scope of what we call "Global Change". This is the background to developing and offering the following course.

The Undergraduate Honors Colloquium: a Semester Long Course

The undergraduate Honors Colloquium on Global Environmental Change was one of a series of colloquia that are offered to honors students yearly. Normally students in their second year take a colloquium. However, this particular course had 40 students, ranging from freshmen to seniors, from a variety of disciplines. The expectation was that the students would leave the program better prepared to come to grips with national and international global change concerns. An important component of the course was the evening public lectures by invited experts. These lectures were given on 12 Wednesday evenings during the semester and were attended by the public and the university community, as this colloquium was a real effort in community-based learning.

The course was team-taught by Margaret Leinen (now Dean of the Graduate School of Oceanography) and myself. We organized the complex science and social issues into manageable units for study. While building each unit, we also developed and exposed students to the extent of interdisciplinary understanding, as well as to trade-offs and compromise that would be required to deal with these immense issues. Students left the colloquium with the following four skills related to global change:

- basic literacy in the physical scientific and social complexity of the problem
- recognition that scientific conclusions are based on an interpretation of data which can produce uncertainty
- skills needed to understand and evaluate consequences and policy options based on scientific projections
- the ability to ask the "right" questions of the scientific community, especially under uncertainty

The colloquium consisted of the public evening lectures by visiting experts plus two class meetings per week for the semester. One class meeting was a lecture; one a discussion. The topics were divided into three major sections and offered in order: Science Issues, Climate and Society, and Policy Options and Mechanisms (for specifics see lecture topics and speakers). The physical sciences were addressed by an overview of the entire topic and by discussions of specific problems, such as the greenhouse effect, ozone loss, and the role of the oceans. This scientific background set the stage for: 1) a definition of potential global environmental problems; 2) a discussion of the methods used to study it; and 3) an evaluation of the uncertainty in the scientific data and conclusions.

Climate and society issues were addressed next from two perspectives: anthropogenic activities and likely impacts, including how we learn about issues. The role of the press began this section. Other topics included an overview and discussion of the population growth, energy use, and technological development that have led us to the present environmental situation. An in-depth discussion of resource distribution issues included: food and fiber, energy, biodiversity, and tropical forests. A final consideration included the potential health effects of projected global environmental change.

The final section focused on policy options and mechanisms. Some of the overall questions addressed were:

> What are the economic issues and the mitigation options that are available?

Do we have the national and international ability to prevent, mitigate, or live with these potentially devastating problems?
How will developing and developed countries differ in their responses to predicted global environmental changes?
How have scientific findings influenced national and international policies?

The exploration of each of these topics included assigned readings, presentations by URI experts, public presentations by visiting speakers, in-class discussions with the visiting speakers, critiques of presentations, reviews of additional literature, and three written integration papers (one per major topic). At the time of this course there was no appropriate text, so the students read papers from the literature. Since then the bibliography for this course supported the development of a *Book of Readings in Global Change* (University of Maryland, International University Consortium). Since some of the information will change rather rapidly, the book will not be bound and will be subject to periodic updates.

In addition to a final essay exam and the other short discussion papers mentioned, the students were required to do a project. They were encouraged to do projects that would educate others in the community about global change issues and to develop interdisciplinary teams. Some of the projects undertaken included the following: a study of the university recycling program and identification of the problem areas; a special insert in the student newspaper on global change issues; a series of taped interviews with other students on campus to identify environmental attitudes held; development of an advertising campaign to alert community members to the problems of non-collection of CFC's during refrigerator and air conditioning recharge; a study of the implications of global environmental change predictions for Malaysia; and an energy audit of portions of the University campus.

The *process* that was undertaken in this course was just as important as the content in the final outcome. The students left this course with a better appreciation of not only the issues of global change, but also the importance of considering the issues from a variety of perspectives, such as those of future scientists, journalists, engineers, teachers, historians, industrialists, and citizens. They also became empowered by understanding the role that they can play as participants in environmental policy-making.

From the very beginning, it was crucial to require student interaction in discussions, as well as in a team project. While learning begins with awareness and knowledge, education in changing long-term behavior usually evolves through steps that can include comprehension, development of skills, application, analysis, synthesis, participation, and evaluation. Knowledge alone is isolated, comfortable, and often of a short-term duration. Students like to know they have the *right* answer. In addressing these complex issues there are no *right* answers; there are perspectives and trade-offs. Moving from the safe but isolated circle of knowledge to any of the other steps requires an output of energy and risk-taking. In order for students to be able to increase their level of risk, a safe, trusting environment had to be developed that encouraged interactions from them that involved something about what they really thought and not what they thought the *right* answer was. Their 'comfort level' with such interactions increased dramatically as the semester went on. It was important to keep emphasizing that the point of the course was to encourage what people should be thinking *about* and not what they should be thinking.

One cannot assume that people automatically move from knowledge to synthesis, or through any of the other steps. All too often the specifics of knowledge are lost after the final exam. Since the real point of education is to change human behavior, the process by which people are encouraged to take more risks by moving from a knowledge base to an ability to evaluate or synthesize is equally as important as the knowledge base itself. Without the forced interaction, the need to learn to communicate with one another in a meaningful way, and the non-judgmental class atmosphere, the educational process would not likely have been so successful. This course is still being cited at URI as a premier example of an important cross-disciplinary and interdisciplinary learning experience[1] .

---

1 A number of educational products and efforts have been generated by the course:

- *Video taped interviews* of half-hour duration each were made with the visiting experts. These video taped interviews are available for courses and educational purposes.
- The videotaped interviews are also the basis for a *distance-learning course* developed by the University of Maryland, International University Consortium (IUC).
- To support the distance-learning course, IUC has developed a *course guide* and a *Book of Readings* based on the colloquium bibliography.
- Two special issues of the Marine Technology Society Journal on *Global Change* have been developed to extend the concept of educating the community and the public.
- My office is responsible for developing a *graduate overview for teachers*, based on the colloquium. We hope this course will be offered during the school year '94-'95.
- I would be happy to share my course syllabus and bibliography.

Building an Undergraduate Earth Systems Science Education Program

Jeffrey J McDonnell †
College of Environmental Science and Forestry
State University of New York
Syracuse
New York
USA 13210

Introduction

This paper describes an undergraduate teaching program based on integrated Earth Systems Science (ESS). Although the view of the Earth as an integrated system is not new, teaching ESS as an independent interdisciplinary subject combining the essential interfaces between physical, chemical, and biologic processes is new, at least in the U.S. university scene. The motivation for developing the aforementioned program resulted from involvement in the National Aeronautic and Space Administration (NASA) and Universities Space Research Association (URSA) Earth System Science Education (ESSE) Program (Fisher, 1994; Johnson and Kalb, 1994). This program was initiated in 1991 as a five-year pilot effort to bring together university teachers and students, along with NASA scientists, to teach the fundamentals of global change science.

The program described herein is one that was implemented by the author and colleagues at Utah State University, one of 22 ESSE schools in the U.S.A. A strong motivating factor for initiating this program at Utah State University was the pre-exiting interdisciplinary centers and institutes already on campus (Table 1). The presence of these various centers meant that many of the institutional barriers to interdisciplinary teaching and research had already been overcome. At the teaching level, an existing and active degree-granting Watershed Science Unit was an ideal model for the facilitation of faculty teaching in departments other than their own.

The following is an overview of the ESSE program at Utah State University, as of June 1993:

† formerly, Watershed Science Unit, Utah State University, Logan, Utah, USA 84322

NATO ASI Series, Vol. I 29
Global Environmental Change in Science:
Education and Training
Edited by David Waddington

Freshman Level Teaching

Within the scope of the ESSE program, the objectives for implementation of the freshmen level course were to (1) provide an overview of the earth as an integrated system, (2) motivate and stimulate interest in ESS issues, and (3) stress the interrelationships between scientific issues and their social, economic, and political contexts. In order to facilitate this broad and interdisciplinary form of teaching, faculty from Watershed Science, Forest Resources, Civil and Environmental Engineering, Plant Soil Science and Biometeorology, Geology and Geography, and Earth Resources were involved in course lecturing and material preparation. Weekly coordination meetings were arranged among six faculty members in order to coordinate and discuss interdisciplinary teaching strategies. In addition to standard lectures, several special teaching tools and techniques were used on an experimental basis. These included the use of 'toy models' (eg., SimEarth™), different video series (eg., Planet Earth) and role playing exercises.

CENTER	AFFILIATED COLLEGE
Center for Atmospheric and Space Sciences	Engineering
Space Dynamics Laboratory	Engineering
Ecology Center	Natural Resources
Utah Water Research Laboratory	Engineering, Natural Resources, Science
Institute for Land Rehabilitation	Natural Resources

Table 1 Interdisciplinary centers on campus that helped in ESSE implementation.

Table 2 lists the course outline used in the freshman-level course. Considerable debate on course content occurred among the faculty. This was based on departmental affiliations, interests, perspectives, and biases. The breakdown of course material into the different sections represented the best 'compromise' for getting across the necessary information in a fixed amount of time (i.e., one ten-week quarter). After an introduction to the concept of the earth as a system and an overview of the NASA interest in ESS educational issues, the basic processes surrounding the earth system were presented. The topics shown in Table 2 were not selected as representing an exhaustive list of key processes, but rather as a number of essential points that would be followed up later in the course during discussions of interdisciplinary linkages. After covering the basic material, linkages between climate and hydrology, biogeochemical cycling, and other coupled processes were discussed. This portion of the course was seen as the innovative aspect of the teaching during which the overlap between traditional disciplines was explored. Following a treatment of systems function and theory, the various tools and approaches to ESS research were examined. Remote sensing factored heavily into course materials in this section as a result of faculty interest and NASA involvement in the course. Finally, the human dimension of global change was woven into a previous discussion of basic physical and chemical processes by highlighting a number of global change issues, including acid deposition, population growth, decertification, and climate warming.

SECTION	LECTURE TOPIC
Introduction	*Introduction to Earth System Sciences* *What is Global Change?*
Solid Earth Processes	*Solid Earth Basics* *Geologic Time and Rates of Processes* *Focus on the Cenozoic* *Carbon* *Radiation* *Atmospheric Circulation* *Climate Dynamics* *General Circulation Models* *Hydrologic Cycle*
Linkages	*Climate-Hydrologic Interactions* *Chaos* *Climate Modification* *Biosphere Components* *Systems Approaches*
Tools, Approaches, and Summary	*Remote Sensing and GIS* *NASA's Earth Observing Program* *Global Climate Modeling* *Highly Visible Geophysical Problems* *Social Economic and Political Concerns*

Table 2 **Survey-level course outline**

Although successful overall, an number of problems were noted with this course. The team-teaching approach, while coordinated through weekly meetings, presented a problem to students more familiar with single instructor classes. Furthermore, individual team members had difficulty presenting difficult and technically-rich scientific concepts in a jargon-free and equation-limited format. Students within the class came from over 12 different disciplines, including many from the humanities, sciences, and natural and social sciences. Issues of breadth versus depth were cited on evaluations by students, but with no consistent argument for one treatment over the other.

Senior Level Teaching

A senior level course was developed around a more focused portion of the Earth System, namely Climate-Hydrologic Interactions. The student group comprising this course was much more homogeneous and included only students from Watershed Science, Forestry, and Civil and Environmental Engineering. As a result, the course was taught with a high level of scientific comprehension, including the use of differential equations and higher level mathematics. The objectives of the course were to (1) provide an in-depth view of climate-hydrologic interactions, (2) show how land-atmosphere moisture exchanges influence ecological and ecophysicological processes, and (3) examine hydrologic issues relating to global change, the role of measurements and modeling strategies.

The course was structured around a visiting seminar series (Table 3). Guest speakers were identified from colleagues at Utah State and at other ESSE schools and NASA centers. The invited lecture slots provided an overview of research topics in Climate-Hydrologic Interactions and was the basis for a lab series and a series of tutorials with individual students on topics covered by visiting speakers. While accommodating students in class, visiting speakers also provided a vehicle for exposing the larger university community to global change research and earth systems teaching philosophy.

WEEK	LECTURE TOPIC	LECTURER
1.	The USU-NASA Earth System Education Program	ESSE Faculty
2.	Evapotransipiration and the Importance of Scale	ESSE Faculty
3.	Hydrologic Water Balance Modeling	ESSE Faculty
4.	Regional Scale Surface Moisture	ESSE Faculty
5.	Paleo Flood Hydrology	Invited ESSE school
6.	Global Climate Modeling with CCMI	Invited NASA
7.	Sea Level Variations and Global Change	Invited Non-ESSE School
8.	Climate-Hydrologic Interactions in the Geologic Record	USU Faculty
9.	Human Response to Local and Regional Climate Change	ESSE Faculty
10.	Water Resources and Global Change	Invited ESSE School

Table 3 Senior course outline (weekly lecture topics)

Several special hands-on tools were used in the senior class. Firstly, STELLA™ modeling software was the basis for the laboratory component of the course (see Fisher, 1994). Table 4 shows the lab syllabi which focused on model building and model analysis of climate-hydrological linkages. The STELLA™ approach enabled students to become involved in model building and sensitivity analysis in a highly focussed, hands-on way. STELLA™ exercises built upon students' knowledge of the software and technical details of climate-hydrologic processes, and culminated in an independently developed model/exercise. Overall, this portion of the course worked very well.

LAB NUMBER	LAB TITLE
1.	STELLA Basics and System Modeling Concepts
2.	Lake Water Balance in a Semi-Arid Closed Basin
3.	Principle of the Earth Energy System
4.	Seasonal Insolation, Surface Cover, and Surface Temperatures
5.	Seasonal Dynamics of Simplified Snow Covers
6.	Climate, Food Production, and Population Growth

Table 4 STELLA lab course outline

The invited visitors from other ESSE schools and NASA centers were an essential part of the senior level course and the Utah program. In addition to lecturing on topics related to their own research, visitors were used to lobby the administration on the benefits and importance of ESSE. Visitors filled in gaps in the local technical expertise for course teaching, and motivated and stimulated the interest of faculty and students in the program. As one of the ESSE schools, a NASA "mentor" was identified in order to help facilitate the flow of data and instructional material from NASA to the university for both the senior-level and freshman courses. This mentor visited the university and spoke to the senior class and participating faculty on NASA's perspective on ESSE.

Summary and Programmatic Issues

Overall, the ESSE program was, and continues to be a success. Nevertheless, several problems associated with organizing interdisciplinary teaching plagued the first two years of course implementation, most notably those associated with difficulties in large team-taught classes. As the program continues, several ongoing debates regarding course structure, content and implementation continue, including:

- What common scientific underpinning is needed to link broad interdisciplinary earth system education?
- How should the human dimensions of global change be covered, including economic, political, and social issues?
- How do faculty receive credit for teaching in departments outside of their home department?
- How does one insert enough flexibility into professional programs, like civil engineering, forestry, etc., so that students can enroll in ESSE classes?
- What level of mathematics should be expected of students at the undergraduate level, when the student body comprises both humanities and science?

Resolution of these issues at Utah State and other schools implementing this style of interdisciplinary teaching will greatly improve the chances of earth systems science becoming a viable and important undergraduate teaching endeavor.

References

Fisher G (1994) This volume

Johnson D and M Kalb (1994) The NASA-USRA university-based program in Earth System Science Education, Paper Presented at the 3rd Symposium on Education, American Meteorological Society Annual Meeting, Boston, Mass., pp. 38-41

Training atmospheric scientists : examples from Denmark

Henrik Skov and Jes Fenger
National Environmental Research Institute
Department of Emissions and Air Pollution
Frederiksborgvej 399
DK-4000 Roskilde
Denmark

Denmark is a small country with a population of 5 million and thus the amount of money spent on environmental issues is limited. This encourages universities and research institutions to work together and this paper illustrates one such collaboration. This is being done within the framework of The Environmental Research Program. Furthermore, a Ph.D. net-work between several research institutions and universities is being planned.

Denmark has 6 universities. Three of these are performing research in atmospheric chemistry in cooperation with two national institutions - Riso and the National Environmental Research Institute (NERI). Our work is based at Odense University and NERI.

The education of students up to Ph.D. level with special emphasis on global change will be presented in this paper. The purpose of the education is to obtain qualified atmospheric chemists with a high scientific standard and with experience in the field of global change.

Undergraduate students

First and second year students, at Odense, can choose a 30-hour course in Air Pollution. The aim is to give a qualitative introduction to important aspects of air pollution.

Although many books are written on air pollution none of them cover special Danish aspects at university level. Therefore, scientists at the National Environmental Research Institute, Copenhagen University and The Danish Technical University are collaborating to write a new textbook. The Danish title is translated to "Air Pollution". The contents of the book are:

NATO ASI Series, Vol. I 29
Global Environmental Change in Science:
Education and Training
Edited by David Waddington

Background
Chapter 1 Formulation of the problem
Chapter 2 Sources, emissions and levels of pollution

Processes in the atmosphere
Chapter 3 Physics
Chapter 4 Chemistry
Chapter 5 Deposition

Effects
Chapter 6 Ecology
Chapter 7 Health and comfort
Chapter 8 Materials

Sources
Chapter 9 Stationary sources
Chapter 10 Purification technology for stationary sources
Chapter 11 Mobile sources

Analytical techniques
Chapter 12 Emission
Chapter 13 Absorption and deposition
Chapter 14 Models of air pollution

Economics and Legislation
Chapter 15 Socio-economics
Chapter 16 Danish legislation
Chapter 17 International conventions

From the contents of the book it is seen that the course covers a wide range of subjects and students are introduced to many subjects and disciplines important for the understanding of atmospheric pollution.

There is also a course for third year students at Odense concerning Environmental Chemistry (the chemistry of soil, water and air). The aim of the course is to give the students a fundamental understanding of environmental problems, so for example students employed afterwards by industrial companies can take environmental issues into consideration in planning. Furthermore, this course is compulsory for students who want to specialize in atmospheric chemistry.

The course concentrates on Danish problems. However, to obtain a full understanding of air pollution, it is necessary to include global change issues like, for example, regional scale transport of pollutants from Central Europe to the Scandinavian countries and their effect here and the decreasing ozone concentrations in the stratosphere over the northern hemisphere. The textbook used is Environmental Chemistry (Manahan 1991) which covers most subjects in the course.

Bachelor projects

The Bachelor projects extend to a 250-hour study which includes a literature survey as well as laboratory work. It is assessed by a report. An example of a project is the construction of a map of a pollutant in the vicinity of Odense City. For example, the concentrations of nitrogen dioxide in the air has in several cases been studied by passive sampling (Palmes tubes). In order to give an adequate description of the importance of nitrogen dioxide in the atmosphere, the students have to understand and discuss for example the interrelationships of those gases such as the various oxides of nitrogen and ozone in the troposphere.

The Master thesis

The Master thesis is a 1 year project. The theses in atmospheric chemistry are mostly carried out at the Joint Research Centre, Ispra site, of the Commission of the European Union (JRC) or at NERI.

At the moment there are two MSc students at NERI. One student is measuring the concentrations of carbonyl species, and the other one ammonia, in the troposphere. At JRC, another student is studying the photochemical degradation of dimethyl sulphide (DMS) in the laboratory which is important for a possible global climate feet-back mechanism (Charlson et al 1987).

Among other subjects investigated, results are the photooxidation of HCFCs, the fate of formaldehyde, and the fate of alkenes and terpenes, all in the troposphere. To get a sufficient knowledge in atmospheric chemistry, all the students have studied Atmospheric Chemistry (Finlayson-Pitts and Pitts, 1986) parallel with the experimental work.

One point to be highlighted is that it is a very good idea to have students working outside the university before they study for the master degree. It has given them excellent opportunities to make the acquaintances of other research institutions and to work on subjects and with equipment which otherwise would have been impossible. Finally, students start on their Ph.D. study at an institution they already know and have a much faster start than usual, obtaining interesting results at a high scientific level.

Doctoral studies

Doctoral studies in Denmark have recently been reorganised. In the old system, the students confined their study to their project and did ad hoc courses. In the new system, students will take 300 hours of courses and have to choose from a wide range of courses. Thus, we hope that students will undertake a more multidisciplinary education and gain an understanding of topics other than their own.

Among Ph.D. projects studied so far the following examples are given.

- At JRC, the photooxidation of dimethyl sulphide and the nucleation of the photochemical products to form cloud condensation nuclei; the night time oxidation of isoprene initiated by NO_3.
- At NERI, the development of a three dimensional hemispheric air pollution model used for the Arctic with special attention towards the sulphur species of both natural and anthropogenical origin; the development of methods to measure ammonia concentrations and fluxes and hydrogen peroxide.

The projects cover disciplines like global climate change, changing photochemical activities in the troposphere and acidification together with autrofication.

Most projects in atmospheric chemistry performed so far have a multidisciplinary nature. For example the study of the role of DMS on the global climate on earth has involved physical chemistry; organic and inorganic synthesis; particle and cloud physics and analytical chemistry. This example illustrates well one of the key points of this Workshop; the multidisciplinary nature of global change science. It is, of course, impossible to obtain a high scientific standard in all these disciplines, and the aim of the education has been to reach a high scientific level in one of the disciplines and a basic knowledge in the others in order to be able to communicate across the professional barriers.

References

Charlson RJ, Lovelock JE, Andrea MO and Warren SG (1987) Oceanic phytoplankton, atmospheric sulphur, cloud albedo and climate *Nature* 326, 655-661

Fenger J and Tjell Chr. In preparation (1994) LuftForurening. Polyteknisk Forlag Denmark

Finlayson-Pitts BJ and Pitts JN Jr (1986) Atmospheric Chemistry. John Wiley & Sons. New York

Manahan SE (1991) Environmental Chemistry, 5 ed. Willard Grant Press, Boston

The Arctic Studies Program at the Arctic Centre, University of Lapland

John C Moore and Manfred A Lange
Arctic Centre
University of Lapland
Box 122
Fin 96101 Rovamemi
Finland

Introduction

The Arctic Centre (AC) is a relatively new institution at a young university. The recent trend of studying peripheral areas, such as the Arctic, actually in the region is a welcome and interesting development in education (see Lange 1994 for a discussion of the benefits of such developments). The AC was set up to provide several roles: research into the impacts of change in the Arctic, an information and data service for the scientific community and for decision makers in the political and private sectors, and a science centre exhibition for the public; and to provide an education in the many facets of the Arctic for students from all over the world via The Arctic Studies Program (ASP). The ASP was first introduced in the spring of 1992, with the first full year program run in the academic year 1993-4. Because the ASP is designed as an international course, all tuition is in English.

The ASP is an undergraduate course that aims to instil an interdisciplinary approach to the problems of the Arctic. The course content covers a wide range of disciplines from the humanities, social sciences, and natural sciences. Also, the students come from a wide range of undergraduate and postgraduate degree courses (Figure 1), and from an increasingly wide range of countries (Figure 2). Rather than equipping conventionally trained specialists (whether scientists, lawyers, or economists) with a global (or Arctic) multidisciplinary outlook, the ASP seeks to change the traditional mind-set of the specialist student into a much freer and more rounded one. The students themselves certainly appreciate this approach to the course (Frame, in press: results of student evaluation of the ASP), and the chance to broaden their horizons. In many ways the wide ranging and interrelated issues that they tackle in the course provide a much more realistic training for future careers and lives after leaving university.

The ASP is therefore in some ways an alternative to a masters course, especially for students who may not wish to follow an academic career - those students

NATO ASI Series, Vol. I 29
Global Environmental Change in Science:
Education and Training
Edited by David Waddington

who will play important roles as future law makers, industrialists, engineers, and decision makers. Thus in practical terms, courses like the ASP are of value in providing a broad multidisciplinary approach for students who will hopefully shape a more enlightened development of future society.

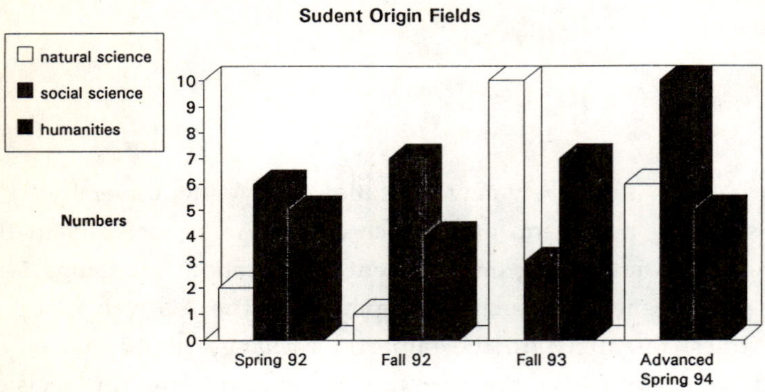

Figure 1. Breakdown of ASP students by field of their degrees in the first 3 ASP Introductory phases and the Spring 1994 Advanced Phase. The natural sciences include physics, biology, botany, environmental science, and geography; social science includes anthropology, economics, sociology; humanities include law, history, international relations, and education.

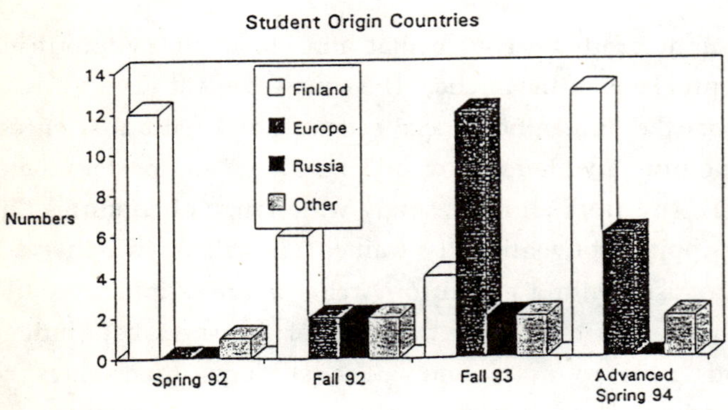

Figure 2. Breakdown of ASP students by their country of origin in the first 3 ASP Introductory phases and the Spring 1994 Advanced Phase. Europe includes EU and Eastern European countries; others include Canada, U.S.A., and Venezuela.

Course Structure and Integration in the Arctic

The ASP consists of two semesters: the Introductory Phase running from September to Christmas (Appendix A), and the Advanced Phase from February to June. The Introductory Phase is structured as a set of courses provided by lecturers in a wide range of topics. Many of the topics address some aspects of global change, whether it be the more directly obvious, such as the role of the Arctic in the global system or the ice sheets and paleoclimate, or less obvious topics (as in some courses in the social sciences and humanities) on aspects of development in the Arctic, such as the environmental and cultural issues involved in hydrocarbon exploitation in the Russian Arctic. The interrelatedness of the issues in the real world is a constant theme and is illustrated at the earliest stage of the course during a week long field trip to the Kilpisjarvi Biological Research Station in the far north of Lapland. There the students sample, often for the first time, such Finnish cultural experiences as the sauna, walk the fells, and meet Sami people in their homes; in addition to receiving lectures on Sami culture and ecology. Back in Rovaniemi, the balance of the course is approximately equally divided between natural science, social science, and humanities, with typically 5 hours of lectures and seminars per day. Students are also expected to study a language (Finnish, Sami, or Russian) or to take a literature assignment. Assessment is based on overall performance in the course, based on student input to individual lecture courses, a field report of the initial orientation exercise in Kilpisjarvi, and a final paper on any subject that is expected to highlight the need and benefits of interdisciplinary approaches to Arctic issues. Topics in the past have included sustainable development in the Arctic, lemmings, and future economic development on the Kola Peninsula.

The Advanced Phase has a rather different structure with a range of short focused courses, usually lasting a week, being offered to students who typically choose 5 to make up their study plan. The courses on offer in 1994 were "The Arctic and the Global System", "Arctic Ecology and Management", "The Baltic Sea Environment", "Winter Ecology and Physiology", Managing Biological Resources", "Military and Environment", "Legal Rights and Cultures of Minorities", and "Art, Culture and Design". Additionally an important part of the course is in the form of field trips such as to the Kola Science Centre and to the Yamal Peninsula in Siberia, and participation in international conference such as the Calotte Academy this May in Inari (Finland), Svanvik (Norway), and Nikel (Russia).

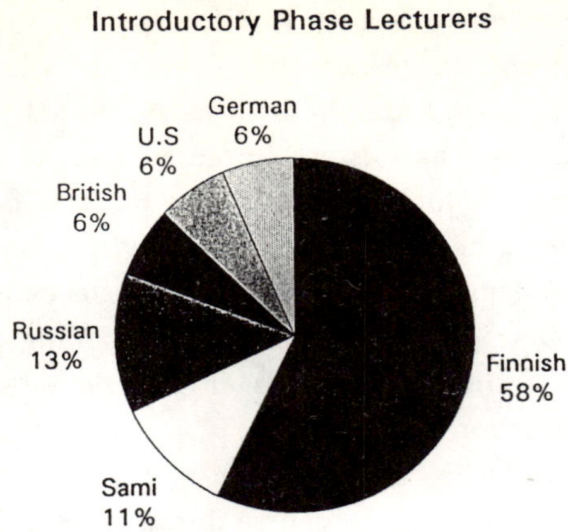

Figure 3. Breakdown of country of origin of ASP Introductory Phase teaching staff.

From the start, the national and international cooperation has been an important part of the ASP with teachers and experts coming from many countries (Figure 3): from institutes such as the University of Aberdeen, the Nordic Sami Institute (Kautokeino, Norway), and the Kola Science Centre; and from many in Finland (e.g., University of Oulu). (See Heininen 1994 for a fuller discussion of the mechanics and cooperation of the ASP.) A significant disadvantage of the ASP and probably of other undergraduate courses is the lack of opportunity to develop research projects that would exist as the basis for student thesis work. The Arctic Centre research section would like to address this problem by bringing in graduate students to do research in the AC, leading to either masters or doctoral degrees, or to a one year diploma. The Advanced Phase of the ASP requires students to complete a 15 page report that, while providing a challenge to undergraduates, is usually limited by time and other course commitments to fairly straightforward literature reviews and field work accounts. The future development of this research aspect of the ASP is aided by the AC proximity to, and good relations with, the Kola Science Centre (KSC) in the Murmansk Oblast of Russia. The KSC is the largest research institute in the Circumpolar Arctic and can provide numerous research opportunities through cooperation and student exchange. (Several KSC employees have participated in the ASP as students and lecturers.) The direct experience of the Kola Peninsula with its severe economic and environmental problems provides the ASP students with firsthand experience of the problems of Russia in general, and exemplify the problems of development in the Arctic.

Global Environmental Change Aspects of the ASP

For a number of reasons the polar regions attract ever increasing attention within the study of global change (see below). This fact should be reflected in the curriculum of a course dealing with the Arctic. We do this by presenting global changes in the Arctic as one of the overarching themes in the Introductory Phase, as well as in a focused short course of the Advanced Phase. In so doing, we stress the interconnectedness of the components of the Arctic system on one hand, and the linkages between the Arctic and the global system on the other. This theme also lends itself nicely to addressing the feedback between the natural processes and the social/human dimension issues in the Arctic. By addressing global change issues in the ASP, we also have the opportunity to deal with general issues of international research programs in the Arctic and their integration into other global change programs. Finally, we also cover questions relating to political decision making, as applied to policies that seek to avoid, or at least mitigate, global environmental changes in the Arctic. More specifically, we address the role of the Arctic in controlling the earth's climate, as a repository of climate information from the past that is available from ice cores, and the impact of industrial development in the region. The Introductory Phase, as may be expected, relies rather more on lectures and fairly direct teaching than does the Advanced Phase course, "The Role of the Arctic in the Global System", where students are expected to make much more active contributions during seminars and role playing exercises.

Dealing first with the paleoclimatic record, the personal experience of one of this paper's authors in the recent deep drilling program in central Greenland (GRIP) is used to present both the scientific findings regarding the pronounced variability of the climate in the past, and also the more personal impressions of being involved in a large scale international scientific collaboration. This combination is designed to try and instil the sense of excitement that can accompany real scientific research. One of the things that is not neglected in the course is the justification for the damage to the environment that necessarily accompanies such a project in the Arctic, and the political priorities that led to GRIP being funded, while other perhaps equally deserving science was not. Another aspect is the often lively debates that ensue during the course on the limits of current scientific knowledge and also on the limits of the imperfect scientific knowledge of the past and present, in modelling the future, and consequently the limits of science in political decision making.

The second role of the Arctic as a major player in the global climate system is covered with extensive reference to the current high profile of the greenhouse effect and the ozone hole. The IPCC reports on future climate change (Houghton et al., 1990, 1992) illustrate the very large changes in temperature that are expected to occur in the Arctic as a result of anthropogenic greenhouse gas emissions. The temperature rises predicted from GCM models are of the order of 3°C over the next 50 years. Such rapid changes exceed any changes in the historical record of climate, and almost certainly exceed any changes that have occurred since the end of the last ice age. The ice core records suggest that the last 8000 years of climate history have been remarkably stable compared with the previous 100,000 years at least. The reasons for such stability and the causes of its breakdown are under keen investigation at present, but it appears that changes in oceanic circulation play an important role. A majority of the sea ice in the Arctic Ocean spends the summer at temperatures within 2°C of the melting point, a rise in temperature would lead to a rapid decrease in sea ice cover in the Arctic Ocean as the positive feedback effect of lowering albedo by removing sea ice allows more heat to be absorbed causing more sea ice melting. Thus, it can be imagined that rather dramatic changes could take place within a very short period of time in one of the climate system's main parts.

The third aspect of the Arctic in environmental change concerns the human impact of industrial development in the Arctic. This is particularly obvious in the marine pollution of the Arctic Ocean with significant input of radiation around Murmansk and Novaya Zemlya, and large, very seasonal (due to freezing) inputs of PCBs and heavy metals from the main rivers of Russia into the Arctic Basin. The recent changes in the political situation in Russia have allowed investigation of the environmental damage around the large metal smelting plants in the Kola Peninsula to be internationally monitored. Damage to the ecosystems in such a fragile environment are likely to take generations to recover, if ever. Several research projects at the AC are studying the effect of air and water pollution on trees and sea ice organisms. It is, however, very striking that the attitudes towards environmental damage that are characteristic of most westerners are not shared by the Russian inhabitants of the Kola peninsula, most of whom are very happy to have well-paid employment in the mines, factories, and nuclear power plants, instead of doing the same job for half the pay in middle Russia (if they can get a job at all). While there is concern for the future state of the environment, the more everyday problems associated with providing themselves and their families with adequate food and other necessities mean that problems that may appear in 20 or 30 years time have low priority. This factor is a very important one to consider, and contrasts with the standard element of academic courses in which students are usually only presented with the results of the damaged environment. It is something that is best grasped during a visit to the region. The fragile nature of the Arctic environment is also a key element in studies of the impact of climate change in the region. Most of the Arctic plant life is limited not by climate, but by nutrient availability. The limitations of the often simplistic assumptions used by models of changes on plant species limits and agricultural practices resulting from climate changes, do not take into account other important factors such as the reduction in biodiversity that will arise as less cold hardy plants out-compete the traditional Arctic plant species.

One further aspect that we like to cover is the impact of change on the indigenous peoples of the Arctic. This again can be illustrated in Russia where there is currently interest in developing the Northern Sea Route (the North-East passage) to serve the Siberian Arctic Ocean ports and function as a trade route from Western Europe to the Far East. This could become much easier in future if climate warming reduces the Arctic Sea ice cover (a factor which is very sensitive to rises of a few degrees in temperature). Increased commercial traffic and pollution in the Arctic would clearly have impact on the traditional fishing and hunting activities of the indigenous people, and the possibility of much easier communication with the region would also bring changes to their way of life.

Assessment of the individual courses of the ASP depends on the requirements of the lecturers and is therefore quite flexible. For the Advanced Phase course on the Arctic in the global system, assessment was based on a 15 minute seminar given by each student on the last day of the course. (Topics included "Natural Climate Change", "Greenhouse Gases: Sources, Sinks and Trends", "Sea Level in a Warmed World", "The Role of Sea Ice in the Climate System", "Changes in Cereal Agriculture in a Greenhouse World".) Secondly, in the 2 weeks following the course, students produced an essay of 3-5 pages in length that emphasised the interdisciplinary nature of decision making in environmental issues and required the student to express a personal but well-founded view. This is illustrated by the following selection of essay topics:

"Explain the possible benefits of global warming in the polar regions and weigh them against the adverse effects of warming. What would be the balance between benefits and adverse effects in your view?"

"Suppose you were asked to make a choice between building a large nuclear power plant or to utilize an alternative form of energy production (hydropower, wind power) in the Arctic. What would be your choice, based on an assessment of relevant environmental factors, and how would you justify it ?"

"Consider that you are a high-ranking politician. You have to make a decision of a) initiating a profitable development project (e.g., exploitation of oil and gas) in an area that is occupied by indigenous people, which would considerably raise the standard of living of these people through shared benefits; or b) of 'leaving them alone', knowing that it will take them a long time before they will ever 'catch-up' with the 'modern world'. What would your decision be and why ?"

In summary then, the Arctic Studies program seeks to provide a rounded education for undergraduate students from across the range of the humanities and the natural and social sciences. Global changes and their relationships to the Arctic are a fundamental part of the course structure and provide one of the main thematic links between the individual courses that make up the ASP structure. Finally, by emphasising the links and feedback between the human and social worlds of the Arctic and the fragile, but still relatively undeveloped natural environment, we hope to leave a lasting influence on the students in whatever career they take up on leaving university.

References

Frame S (In Press) A season of discovery - reflections on the Arctic studies program and a winter in Lapland. Arctic Centre News.

Heininen L (1994) Arctic studies - for academic and practical purposes. In: *Teaching the North: the Challenge of Circumpolar Studies Programs*, L Heininen and L Muller-Wille (eds), Arctic Centre, University of Lapland Rovaniemi, 17 - 19.

Houghton JT, Jenkins GJ and Ephraums JJ (eds) (1990) Climate Change. The IPCC Scientific Assessment, Cambridge University Press, Cambridge.

Houghton JT Callender BA and Varney SK (eds) (1992) Climate Change 1992. The supplementary report to the IPCC Scientific Assessment, Cambridge.

Lange MA (1994) The Arctic in a changing world: new challenges for circumpolar universities. In: *The Changing Circumpolar North: Opportunities for Academic Development*, L Heininen (Ed) Arctic Centre Publ. 6., Rovaniemi, 151-165.

Appendix A

Arctic Studies Program
A Program for the Introduction to the Arctic

The Arctic Studies Program (ASP) starts with the Introduction to the Arctic (15 study weeks), which provides a broad introduction to the Arctic and its manifold aspects. This introduction aims to broaden students in arctic and northern affairs, to demonstrate the complexity of processes in the region, and to highlight the interrelationship between these processes and others at work around the globe. It also strives to illustrate the interdisciplinary nature of issues relevant to the Arctic or the Circumpolar North and the particular role played by the indigenous peoples living there.

The Introduction to the Arctic includes orientation comprising both field (an excursion to Kilpisjarvi in Finnish Lapland) and classroom work. The basic course consists of several thematic units, seminars, and, if possible, participation in an international seminar or conference. In addition, language courses (through the University Language Centre), literature, and offerings from other departments are available. Course requirements include the research and preparation of a 20 page study paper.

Structure of the program

1.	Orientation	(2 study weeks)
2.	The Arctic in the Global System	(3 study weeks)
3.	Indigenous People of the Circumpolar North	(3 study weeks)
4.	The Arctic as a Distinctive and Multiple-Use Region	(3 study weeks)
5.	Languages or Literature	(2 study weeks)
6.	Study Paper	(2 study weeks)

Academic Fields:

All themes and subthemes of the program are under the following three academic fields:

A. Natural Sciences, "Ecology"
B. Humanities, "Peoples"
C. Social Sciences

The teaching of the program runs either under the three academic fields or be interdisciplinary teaching. Interdisciplinary teaching will mostly concentrate on a relationship between ecology and social affairs.

A Weekly Class Schedule:

Teaching from each of the academic fields during the ten week study period will be done three days each week from Tuesday to Thursday. The teaching will include both lecturing and, once a week, an interdisciplinary seminar or discussion, which deals with a relationship between topics of two or three academic fields. On the 'rest days' of the week (Monday and Friday), there will be languages, individual studying of the literature, and a study paper.

Study Paper:

All students will write a study paper on a chosen topic, which should deal with the themes discussed during the program, emphasizing the interdisciplinary nature of issues and problems of the Circumpolar North. The paper should be based on written material, lectures, and discussions, and not exceed 20 pages.
The paper will be a university exercise, but it should be prepared like a scientific paper, e.g., with complete references. Student work will be supervised by one of the scientific staff members of the Arctic Centre or by one of the key lecturers of the ASP. The idea is that students prepare the paper during the course (by the beginning of December) and discuss it in a seminar in the middle of December.

Assessment System:

The studies of the ASP students will be assessed according to the elements of the course, e.g., a study paper, a report from the Kilpisjarvi excursion, languages, and each student's overall performance (participation in discussion, seminars, group work or other activities, and a seminar of study papers). Details of the assessment system are given to students in September at the beginning of the program.

Examples of some topics on which students are assessed are given in Appendix B.

ARCTIC STUDIES PROGRAM: FALL 1993

GLOBAL CHANGES AND THE ARCTIC: A TAKE-HOME EXERCISE

The following exercise is not so much meant to "test" your knowledge and whether you have been listening carefully to the lectures (I am assuming you have!), but they are rather intended to provide all of you with an opportunity to give me **your** view on issues that we have addressed during the last few days.

Please, do not take more than about ten minutes and do not write more than half a page to answer each question. The whole exercise should really not require more than one hour of your time. However, I would appreciate if you would write clearly, thus helping me to find out what you are saying.

1. What are some of the major processes that link the polar regions with the global system? Identify whether global warming would be enhanced or reduced if changes in these processes would take place in Antarctica or the Arctic.
2. Can you provide (at least) one example each for
   - disturbances on the Arctic system caused by human activities and for
   - adverse effects of global changes on Arctic societies?
3. How would you define the **three top** research questions that need to be addressed in order to better understand the role of the polar regions in global change?
4. Suppose you were a (very powerful) politician. What would be your three top policy actions (e.g., environmental laws) that you would introduce in order to minimise man-induced disturbances on the Arctic system?
5. Returning to our first carbon dioxide exercise (GC1), please provide a cost-benefit analysis for each of the four options in the action/problem matrix. Please note that "costs" in this context goes far beyond the usual monetary connotation and includes things such as "changes in lifestyle", time and labour, losses in personal choice, career options etc.

Please hand your answers in (you can collect them and have them sent to me) no later than November 26.

I am looking forward to your answers. Thanks!

Manfred.

THE ROLE OF THE POLAR REGIONS IN GLOBAL ENVIRONMENTAL CHANGES

TOPICS FOR ESSAYS

Preface

You are asked to pick one of the topics given below for a **three- to four-page** essay. The essay should not so much reflect any "learned knowledge" but rather your personal view. However, it is expected that the latter and what you express in your essays is well founded. So, simple "personal arguments" or any "hand-waving" won't do!

Your essay should be typewritten on A4 paper and should be handed in no later than **April 30** (12 o'clock a.m., sharp!!!). Please, contact either John or myself, if you have any questions.

We are looking forward to getting your essays!

Essay Topics

1. Suppose you were asked to make a choice between building a large nuclear power plant or to utilize an alternative form of energy production (hydropower, wind power) in the **Arctic**. What would be your choice, based on an assessment of relevant environmental factors and how would you justify it?

2. To what extent can/should research on global changes in the polar regions (i.e., Antarctica and the Arctic) be done with identical/similar methodologies and in what fields does it have to be done differently? Please explain and give concrete examples.

3. Explain possible benefits of global warming in the polar regions and weigh them against the adverse effects of warming. What would be the balance between benefits and adverse effects in your view?

4. Give a brief outline of a research project (including the rationale for choosing it) addressing the issue of global changes in the polar regions that you would consider to be of highest priority.

5. Consider to be a high-ranking politician. You have to make a decision of
a) initiating a profitable development project (for instance a project dealing with the exploitation of oil and gas) in an area that is occupied by indigenous people, which would considerably raise the standard of living of these people through shared benefits; or
b) of "leaving them alone", knowing that it will take a long time, before they will ever "catch-up" with the "modern world".
What would your decision be and why?

Education and Training in Environmental Science in Greece

Panayotis A Siskos
Laboratory of Analytical Chemistry
Department of Chemistry
University of Athens
Panepistimiopolis - Kouponia
157 71 Athens, Greece

The first Earth Summit held in Stockholm (1972), warned of the side effects of the growing economic development in industrial countries. At the same time, the European Communities first became involved in the environment protection. The new Treaty of European Unions, 1992, introduced as a principal objective the promotion of sustainable growth which respected the environment (Act 2) (Commission of the European Communities 1992).

The following characteristics of *sustainable development* were identified:

- it maintains the overall quality of life
- it maintains continuing access to nature resources
- it avoids lasting environmental damage.

Today there is growing world-wide concern at the continuing deterioration of the state of the environment and the serious degradation of global life support-systems. The ozone hole, the greenhouse effect, acid precipitation, air pollution, the loss of biodiversity are some of today's global problems of Earth science. The second Global Earth Summit, held in Rio in 1992, has defined the details of Agenda 21 for the protection of Environment (Keating 1993).

Greece, a member of European Union, is facing serious local environment problems and follows the European Environmental policy and legislation. It became clear that the implementation of legislation and the appropriate actions in protecting the environment need well informed people and politicians, trained activists and administrators and the training of scientists with a strong background in environmental issues.

In this paper, I will present the education and training programs in Greece related to environmental science and environment protection in schools and universities.

NATO ASI Series, Vol. I 29
Global Environmental Change in Science:
Education and Training
Edited by David Waddington

Secondary schools

Environmental aspects related to all branches of sciences are being taught in secondary schools in Greece. The book "50 Simple Things Kids Can Do to Save the Earth" (Javna 1990) translated in the Greek language, is given to pupils.

In 1990, the Ministry of Education introduced a law concerning *Environmental Education*. The teachers must emphasise the importance of environmental protection during the teaching of science. In addition, students are encouraged to study local environmental problems.

Undergraduate courses

At the university level, in all departments of sciences, courses (Core or optional) are provided in:

- Environmental chemistry
- Environmental physics
- Environmental biology
- Environmental geology
- Environmental technology
- Environmental education
- Ecology
- Sustainable agriculture

Curriculum requirements and teaching resources for these courses differ from university to university. More specifically, in the department of chemistry at the University of Athens, there are two courses on environmental chemistry. In *General Environmental Chemistry*, the following are studied:

- Basic principles of environmental science
- Water chemistry
- Soil chemistry
- Atmospheric chemistry
- Sources of pollutants
- Effects of pollutants

20% of the course time is devoted to the three global environmental problems (acid rain, ozone layer, greenhouse effect). In addition to the textbook given, a supplementary book, "The Changing Atmosphere: A global challenge" (Firor 1990) translated into Greek, is recommended as a reading resource.

The second course is entitled *Special Environmental Chemistry* and covers:

- Photochemical pollution
- The chemistry of burning fossil fuels
- Cleaner technologies
- Toxic wastes
- Solid wastes
- Liquid wastes
- Recovery and Recycling
- Environmental Analysis:

Sampling of solid, liquid and air. Determination of primary pollutants in air and water. Monitoring techniques.

Postgraduate studies

The coming academic year postgraduate studies will be started in almost all our universities. The traditional system of awarding Master and Ph.D. degrees is changing to the U.S. system of graduate studies. The following long courses, leading to formal qualifications, are given within the core of the program in our department at the University of Athens:

- Advanced Atmospheric Chemistry
- Advanced Water and Marine Chemistry
- Soil Chemistry
- Health effect of air pollution
- Monitoring and analytical methods
- Modern technologies for abatement of pollution

There is also field and laboratory work and research projects.

Short courses for scientists

The Greek Chemical Society has established short courses related to environmental problems, for chemists and other scientists are facing in their industrial work.

Such courses given so are are:

- Environmental microbiology
- Domestic wastes
- Toxic industrial wastes
- Treatment technologies of liquid wastes

Other organisations and scientific societies are also involved in training young scientists and technologists on different topics of the environmental science and technology.

There is a very interesting new development. The National Centre of Public Administration founded in 1983 an educational unit, the Institute of Continuing Training, in order to train on a continuing basis public employees and administrators in new concepts, technologies and legislation related to the practical environmental topics which constitute towards the implementation of the European Environmental Policy.

The following topics have been given the last year in different cities in Greece. The courses are intensive with a teaching period lasting two weeks (100 hours):

- Management of water resources
- Pollution and technologies for the protection of the environment
- Natural environment and natural risks
- Environmental management
- Assessment of environmental impacts.

In all above courses, formal or non-formal, it is not obvious at the first glance the global environmental dimension of the given courses. It is expected that the new training and education concepts, as presented in their workshop will affect the syllabus and the outline of new courses in Greece in the near future.

In addition, the new generation of scientists and activists with an global outlook will affect both the education and practice in facing environmental problems. Public awareness must be another target for educators. Politicians, local authorities and the public should be aware of the global dimension of environmental problems.

References

Birks JW, Calvet JG and Sievers RE, (eds) (1993) The Chemistry of the Atmosphere. Its impact on Global Change : Perspectives and Recommendations. American Chemical Society, Washington D.C.

Commission of the European Communities (1992) Towards Sustainability, Vol I, II and III, Luxemburg

Firor JW (1990) The Changing Atmosphere: A global challenge, Yale University. Translated in Greek by PA Siskos and E Ioannidou. Costrarakis, Athens, Greece 1992

Javna J (1990) 50 Simple things kids can do to save the earth. The Earth Work Group, Berkeley, CA, USA. Translated in Greek by R Branas. Pontiki AE, Athens, Greece 1990

Keating M (1993) The Earth Summit's Agenda for Change: A plain language version of Agenda 21 and the other Rio Agreements. Centre for our Common Future, U.S.A.

Manahan SE (1991) Environmental Chemistry, 5th ed Lewis Publishers Michigan U.S.A.

Siskos PA and Skoullos MJ (1990) Environmental Chemistry, Vol 1 and Vol 2, University of Athens, Athens

Climate Change Education and Training at Wageningen Agricultural University

Rudolf S de Groot
Center for Environment and Climate Studies
Wageningen Agricultural University
PO Box 9101
6700 HB Wageningen, The Netherlands

Introduction

In the Netherlands, Global Change research has a relatively long tradition, especially if we also consider research on acid precipitation and eutrophication as part of the "Global Change-issue". Research on the enhanced greenhouse effect and climate change increased strongly after the establishment of a special national funding structure for this issue in 1990: the National Research Programme on Global Air Pollution and Climate Change.

Initially, both research and education in Global Change at the Wageningen Agricultural University (WAU) focussed mainly on the direct physiological effects of air-(and water-) pollutants and greenhouse gases on plant productivity and cultivation. However, the number of departments involved in climate change research and education has grown rapidly since 1990 with many departments now looking into other aspects including the physical climate system, socioeconomic effects, and policy implications.

In January 1994 Wageningen University started a 5-year multi-disciplinary research programme on the "Greenhouse Effect and Climate Change", coordinated by the newly established Wageningen Institute for Environment and Climate Change Research (WIMEK)[1]. Fifteen departments from both the natural and social sciences are participating in the programme with over 40 projects divided among six main topics: 1) working and monitoring of the global climate system, 2) emission of greenhouse gases, 3) physiological effects on soils and plants, 4) effects on ecosystem stability and biodiversity, 5) consequences for land-use and water management, and 6) socio-economic aspects and solutions.

This research programme is the basis for the education and training provided by the University on the greenhouse effect and climate change. Climate change

---

1 WIMEK was established in 1993 in order to provide the institutional basis for coordinating and developing multi- and inter-disciplinary research on environmental issues at Wageningen University

NATO ASI Series, Vol. I 29
Global Environmental Change in Science:
Education and Training
Edited by David Waddington

education at Wageningen Agricultural University is given at three levels: a) an undergraduate, introductory course, b) an advanced/post graduate course, and c) occasional "masterclasses" and summer-schools. These three education programmes are described in the following sections.

Introductory/Under-graduate Course

Since 1990, the Agricultural University has provided a special course on the greenhouse effect and climate change, which is organised every two years by the Center for Environment and Climate Studies. It consists of a series of lectures (22 in total) which are given over a period of 3 months and divided over the 6 topics of the research programme mentioned earlier.

Advanced/Post-graduate Course

Work is now underway to develop a special course for post-graduate students on "global change". This training course will draw from the experience gained with the under-graduate course, and it is being developed by WIMEK in cooperation with the Institute for Environment Studies (Free University of Amsterdam) and the Open University which is working on an MSc course "Climate and Environment" (Jansen, 1993). The results of this NATO Workshop will be a welcome source of additional information for the development of the course, which is described in some more detail below.

Objectives

The complexity of the climate change issue calls for an integrated, multi-disciplinairy approach not only in research, but also in education and training efforts. Therefore, the intention is to concentrate on the interactions and feedback mechanisms between the various subsystems of the cause-effect-measure chains of global climate change. Incorporated in the course will be discussions of the basic principles of an integrated assessment of global climate change, as well as descriptions of uncertainties, risk analysis, the different perspectives, and policy implications.

The main learning objectives are to enable the student to:

- identify the main (feedback) mechanisms of the global climate system
- analyse some of the present major uncertainties in climate change
- evaluate the implications of the various climate change scenarios,
- develop preventive policies which reduce risks to tolerable levels

Provisional course-contents

The course will consist of four modules (A-D, see box 1). For each module several overview lectures will be given to introduce the main topics, after which the students will have the opportunity to break up in working groups to study the background material in more detail and to apply the obtained knowledge by writing a short essay and/or practicing with integrative computer models.

Box 1: **Provisional contents of post-graduate course on the greenhouse effect and climate change**

A) *Introduction: an integrated systems approach*

A description of the systems-approach and the many perspectives according to which one may describe and evaluate phenomena of global climate change, including scientific, social, and cultural perspectives.

B) *The climate change*

B1. Analysis of the climate system ("how does it work")
- radiation and energy balance, - climate change forcing mechanisms (e.g., solar radiation)
- biochemical feedbacks
- climate change from a geological perspective

B2. Global cycles:
- carbon, nitrogen, phosphorus, sulfur, water,
- linkages between the cycles/synergistic effects (e.g., ozone)

B3. Human impact on the climate system
- resource use (e.g., burning fossil fuels),
- land use (e.g., deforestation, urbanisation)
- agriculture (e.g., methane emission from rice fields),
- Transport

B4. Climate modeling and scenarios
- simple climate modeling (e.g., energy balance climate models)
- General Circulation Models (GCM's)
- model calibration and validation,
- climate predictions, scenarios and their limitations

C) *Impacts of climate change*

Cl. Hydrological effects
- sealevel changes and its consequences
- changes in river discharge/runoff (floodings)
- changes in groundwater level and soil moisture
- changes in glaciers, snow accumulation, snow-melt
- changes in water quality

C2. Land degradation and soil erosion
- soil-erosion and salinisation
- changes in soil organic matter-desertification, coastal erosion

C3. Ecosystem stability
- ecosystem physiology (including direct effects of CO_2),
- ecosystem structure and complexity (effects on species and vegetation, including modeling of zonation and growth),
- landscape-ecological effects (integrated effects on major ecosystem complexes such as forests, magroves, coral reefs)

C4. Land and water use
- cultivation (agriculture, forestry, aquaculture/fishery, etc.)
- water management (including coastal defense)
- nature management and conservation
- recreation
- infrastructure (e.g., transport, urbanisation)
- energy use

C5. Socio economic impacts
- health (heat stress, diseases, water availability/quality),
- food availability (including models),
- extreme events and disasters (storm damage)
- economic costs

D) Implications for planning and policy making

D1. Uncertainties and risks
The integrated climate assessment framework used in this course should represent the reliability with which the various pieces of the climate puzzle are understood, and be able to propagate uncertainties through analysis and reflect them in the conclusions regarding the risks.

D2. Decision analysis
A comprehensive view of past and current global decision initiatives and of the available tools for decision-making is needed in order to evaluate the implications of various climate scenarios and the development of preventive policies that reduce risks.

D3. Integrated Assessment Models
Integrated climate assessment models are being developed (e.g., the TARGETS model by the RIVM), based on the systems approach as described above. This integrated modeling framework makes it possible to evaluate the consequences of several types of human influences simultaneously. It is thereby envisaged that synergystic effects may be brought into view. Much attention is paid to the issues of uncertainty, and how to deal with incomplete knowledge.

Course material

The integrated approach of this course demands a strong structuring of the course with regard to such educational strategies as motivation, creating an active attitude, steering, testing, and giving structural feedback. This complexity of approach is present in the level of integration, as well as in the transition from a global environmental problem to local causes, impacts, and actions. A text-workbook model is therefore proposed, whereby the textbook primarily deals with explanations of the background of the topics involved and the workbook primarily with the educational part of the course.

The **textbook** provides the understanding of the complex interactions and the basic concepts of the integrated assessment approach of the course. After an introduction on the systems-concept, the different aspects of the cause-impact-measure-chain are dealt with in more detail.

The **workbook** will present material that illustrates the contents of the textbook with actual issues and discussions. The main purpose of the workbook is to offer didactical support for students which is, among others, done by underpinning the theoretical concepts and ideas described in the textbook with a number of concrete and practical examples. The workbook also fulfils an important role in the translation from global to local action and impacts (together with the computer model PC Targets). Various integrated assessment studies related to the different scales of impacts of human-induced climate change are dealt with and will form the basis for student assignments. Some selected source material and extended references, that reflect actual problems and discussions related to a chapter or some parts of the textbook, will also be included. These cases or examples may be assisted by pre-produced modeling results.

A **computer model** (PC-TARGETS) forms an integral part of the course. The most important learning objective of the computer model is to apply knowledge of all aspects of climate change from an integrated perspective in a practical way. It will be an important aspect in the translation from global to local action and impacts ("think globally, act locally"). An active selection of socio-economic factors or emission scenarios will be possible on local and global scales. Also, the possible impacts of climate change are assessable on a regional scale.

Student **examinations** will consist of writing an essay based on literature and computer-model results. The learning objectives will be tested through the ability of the student to present and defend his or her scientific point of view.

Occasional "Masterclasses" and Summer-Schools

In December 1993, a so-called "Masterclass" was organised by the Center for Environment and Climate Studies for the first time (de Groot and Ooms, 1994). Similar to Masterclasses in the world of music, this event aimed to provide advanced students and PhD candidates the opportunity to obtain more in-depth information on this new research field and to exchange knowledge and ideas with well-known scientists. A review-committee selected 23 students and PhD candidates from 7 Dutch Universities, basing the selection on a short paper and propositions that the candidates had prepared around three themes: (1) how the climate system works and why it is changing, (2) possible consequences of the enhanced greenhouse effect and climate change for natural and cultivated landscapes, and (3) socioeconomic aspects and policy implications. In a three day period (1-3 December 1993), eleven "Masters" from The Netherlands, the UK, and Norway presented the latest information on their field of expertise in the morning sessions, while the afternoons were reserved for presentations of the propositions by the students and discussion with the masters and the audience. The Masterclass was a great success and may be repeated on a regular basis.

References

de Groot RS and Ooms B (1994) Masterclass challenges young scientists on greenhouse effect, in: "Change" June 1994:16-17, the Research and Policy Newsletter on global change from the Netherlands, Kluwer Ac. Publ.

Jansen D (1993) Course plan of Climate and Environment (24 pp). Open University of the Netherlands, Faculty of Natural Sciences. Heerlen.

Ooms B and de Groot RS (eds), (1994). Reader of the course "Greenhouse effect and climate change" (ca 400 pp, partly in Dutch). Center for Environment and Climate Studies, Wageningen Agricultural University .

Environmental Studies at the University of Colorado

James W C White
Director, Environmental Studies Program
University of Colorado at Boulder
Boulder
Colorado 80309, USA

Introduction

The Environmental Studies Program at the University of Colorado is an interdisciplinary degree program offering a Bachelor's degree (BA) with a major in environmental studies. The major consists of a small number of required courses, as well as a selection of advanced courses in two possible tracks: the Environmental Sciences Track, and the Society and Policy Track. The goal of the program and major is to provide students with a general education in environmental studies with reasonable rigor in either the physical sciences or the social sciences. The degree is structured such that a double major in a few established disciplines can be easily obtained, and that a minor in a discipline is almost assured. The courses are offered in the 17 participating departments. Over 80 courses are offered within the major, taught by over 60 faculty. There are over 500 majors in the program at this time. An internship program is also offered in order to provide students with the opportunity to gain experience in the field as well as to establish contacts in the job marketplace. The Program resides structurally within the Office of Interdisciplinary Studies, not within any host department (Appendix I).

The Environmental Studies Curriculum

The Environmental Studies (ES) curriculum consists of a common set of courses and a choice of two advanced tracks: the Society and Policy Track and the Environmental Sciences Track. Courses making up the common curriculum are introductory sequences in the physical sciences, as well as basic courses in the social sciences. The purpose of the common curriculum is to expose all majors to the physical sciences, and thus the physical workings of the Earth, as well as to the social sciences, and thus to the human dimensions of global change. Sequences in biology, chemistry, geology/geography, and economics are required, along with an environmental ethics course, either through the Philosophy Department or the Geography Department, and a course in public policy through the Department of Political Science. Also part of the common curriculum is one statistics course.

NATO ASI Series, Vol. I 29
Global Environmental Change in Science:
Education and Training
Edited by David Waddington

The Society and Policy Track has three required courses in the areas of human ecology, environmental economics, and conservation practice, and then breaks into four possible specializations. These are Environment and Natural Resources, International Environment and Development, Decision-Making, Planning and Public Policy, and Environmental Analysis. Each specialization has a list of courses from which the student chooses. The primary departments contributing to this track are Geography, Economics, Sociology, and Political Science. Courses are also taught in the Departments of History, Physics, American Indian Studies, Psychology, and Environmental Design.

The Environmental Sciences Track consists of four required courses and three specializations. The required courses are one additional math (a calculus course), and courses in the principles of ecology and basics of hydrology. In addition, a set of two field courses is required. One is an introductory course on field observations and basic techniques; the other is an advanced field course in a choice of topics including, for example, hydrology and structural mapping. The field courses are taught in the Department of Geological Sciences. The three specializations are in water, biogeochemistry, and climate. Each specialization has a number of courses to choose from. The primary participating departments are Geological Sciences, Biology, and Geography. Courses are also taught in the Departments of Chemistry, Physics, and Atmospheric Sciences.

The University of Colorado is on the semester system, and a total of 120 credit hours are required for graduation. Lecture courses are typically 3 credit hours, so a normal load for a student who will graduate in four years is four to five courses per semester. The ES major has requirements totaling about 60 to about 75 credit hours. A typical ES major will take 20 to 25 courses in the major. Thus, it is important that the ES major begin the common curriculum as soon as possible in order to graduate on time. Advising, particularly at the incoming college level, is extremely important to the success of the student and the program, so the ES Program has a full time student advisor, as well as designated faculty advisors in several academic departments.

To complement the curriculum, the Environmental Studies Program offers its majors a number of internship opportunities, in which students earn academic credit while working in selected positions with private and public agencies and firms. The minimum time required is 10 hours per week for 16 weeks to earn three units of course credit. The primary goal of the internship program is to provide academically-supervised opportunities for majors to work in public and private organizations in order to relate theory to practice as an expansion of learning and understanding. Students also get valuable exposure to hiring opportunities. Internship opportunities are available with a wide variety of private and public agencies and firms.

The Environmental Studies Program is only one facet of global change teaching at the University of Colorado. Many of the departments which participate in the ES major offer degrees in areas of global change or offer their majors the option of concentrating in the area of the environment and global change. This situation helps to retain courses which are needed, as the ES major is not the only student served by the classes. In addition, the university has designated one of the freshman dormitories, Williams Village, as the "environmental dorm". Students who participate in the Williams Village Program can take classes in the dorm, some of which also make up the ES common curriculum. These classes have much smaller enrollments than the usual large lecture class and thus offer the student more chance for open discussion. Also offered in the Williams Village Program are discussion courses on aspects of global change, for example, climate change and atmospheric chemistry. These classes, also with small enrollments of 20 or less, provide an excellent discussion-based introduction into global change issues.

Changes to the ES major currently being planned include adding a seminar series which upper level students can take for credit. The seminars will be a mixture of science talks given by faculty working in the general area of environmental studies, and practical advice from persons currently employed in the field. The idea is expose the upper level students to the realities of working in this area. It is clear from interviews with students, that many do not have a good grasp of the job opportunities which await them, nor what they must do in order to qualify for those jobs. For example, while a student pursuing the Environmental Sciences Track many be required to take only one semester of calculus, it is clear that additional math is absolutely necessary for post-graduate work or for employment in many areas. Students are advised to take additional math, but may not see the need unless it is clearly presented to them by potential employers. While the ES major is not a job training program, we are very sensitive to the needs of the students in terms of providing them with employable skills. Another proposed addition to the curriculum are "magnet courses". These are courses which attempt to take advantage of the unique opportunities available at the University of Colorado. For example, the university has strong research programs in the modeling of climate change. A course in General Circulation Modeling (GCM) has been proposed, which will take advantage of that expertise and funnel it to the upper level undergraduate. Other proposed courses include remote sensing and urban air pollution. The University of Colorado has strong research and teaching programs in space studies and remote sensing, and nearby Denver is second only to Los Angeles in the severity of its air pollution problem.

As a final note, it should be pointed out that the Environmental Studies degree replaces an existing program in environmental studies, The Environmental Conservation Program. The changes are far more than in name only. The Environmental Conservation Degree, while very successful at attracting students (in excess of 500 declared majors), lacked scientific rigor and was faulted by the students for not providing a training leading to a job in the field. The Environmental Studies Program seeks to address both of these deficiencies. In particular, we have recently instituted a tracking system of graduates to assess how effective this new degree will be in training students to work in the general area of global change and environmental quality.

Discussion

Teaching in an interdisciplinary degree program, such as Environmental Studies, within a university system which is structured around established disciplines presents some unique challenges. Since the ES Program offers only a few classes itself, and rather relies on classes taught in traditional departments, one of the most important tasks of the Director is to ensure that classes useful to the ES major are being taught at the times needed and with the student spaces required. As rewards for faculty for teaching originate generally within departments, the ES Program has little leverage in accomplishing this task. Departments with large teaching loads and large numbers of majors, for example, see no benefit in adding ES majors to their already full classes. The ES Program benefits greatly from the fact that global change and environmental quality is currently the primary research theme at the University of Colorado, and that it is also a strong theme in many existing departments and hence classes within those departments. Our success in combining the physical and social sciences, for example, can be traced to the high level of enthusiasm at the faculty level to make this unlikely combination happen and work. For a university or college lacking this broad base of faculty support, managing a program like

Environmental Studies would require rearranging the reward structure for faculty, which in turn implies a committed and proactive administration.

The task of ensuring that the necessary courses are taught is easiest at the introductory level, in which the fundamentals of the functioning of the physical and human worlds are taught. The basics must be taught from first principles and understood thoroughly by the student, a concept which applies equally to the basics in the physical and social sciences. Teaching of the basics is already a part of established disciplines, and interdisciplinary programs can take advantage of these introductory courses. The common curriculum of the ES major uses this approach, although students are urged to take advantage of introductory courses which use environmental problems as examples and illustrations. Convincing departments that they should teach introductory courses of direct use to the ES major is therefore not an issue. The task is harder at the upper levels of college, however. ES majors need the information which is taught in upper division courses, but a course which is suitable for majors in a discipline may not be well suited for the ES major. Often, the examples used, the level and focus of detail, and the pace of the course are not appropriate for the ES major. One of the greatest challenges in supplying an interdisciplinary education in environmental studies is fitting all of the courses and concepts deemed necessary into the traditional four year college degree. The struggle of breadth versus depth is a constant one. As the goal is not to produce a shallow generalist, upper level courses become critical to providing the depth of knowledge needed, and the success of the ES program depends largely on these courses. Faculty teaching these courses must either be persuaded to use environmental examples as they teach to traditional majors, or else restructure the course to focus on the environment. The latter approach is becoming more common as students demand (and take) such courses and as the importance of understanding and coping with global change and environmental issues to the future success of society becomes clearer. It is also apparent that the long term stability of the ES program depends in large part on the development and teaching of these upper level courses.

An interdisciplinary degree in environmental studies should be a degree which is useful to the student in terms of gaining employment or in going on to postgraduate study. This sounds obvious, but in fact it is quite easy to lose sight of this goal. Our experience at the University of Colorado has been that we must monitor the rigor of the program, in order to ensure that the needs of the students are being met. The recent restructuring of the Environmental Conservation degree into the Environmental Studies degree is a reflection of this. One strategy which blends disciplinary and interdisciplinary educations is to encourage students to double major with a disciplinary degree. This approach recognizes that established departments and disciplines are dominant in both post-graduate education in the United States and also in many sectors of the job market. Examples of relatively easy double majors with ES include geological sciences, geography, and economics. Other double majors are possible within a four year time period, but require careful planning on the part of the student. Many departments offer a minor, which can also be a method by which students can pursue an interdisciplinary education, but also obtain disciplinary credentials.

Finally, the teaching of environmental studies poses challenges which go beyond the structure of the program and may conflict with traditional disciplinary programs. The central role of humans in global change presents one of the largest challenges, as it requires the merging of social and physical sciences. This is not an easy task. Yet the importance of ensuring success for this marriage is clear. For example, humans are an integral part of all biogeochemical cycles and the climate system. This is not inherently good or bad; we are long past the point at which such value judgments are useful. There is no sense in separating "natural" from "human caused" when looking at biogeochemical cycles and the climate system. Scientists may want to make this distinction in order to understand the systems better, but we are such a major player in the working of these earth systems that the distinction is not useful, and, in fact, gives people the impression that our actions can be cleaned up sufficiently that we would not perturb the "natural" cycles.

It is when we accept our real place in the workings of the earth that we realize that we must incorporate the dynamics of human behavior into the science of global change. The nature of human decision making, from the national and international levels involving politics and social norms, to the individual level involving risk assessment, must be considered, if we are to predict how we humans will act in the future (or for that matter, now). We can perfect our knowledge of the chemistry and physics of the Earth, but this will be of little use if one of the major forcing functions, human behavior, remains unpredictable.

This broad interdisciplinary nature of global change and environmental studies is one of the larger challenges in teaching global change. Classes that originate in science departments do not, in general, deal with the sociological, political, philosophical, or historical aspects of global change. These issues constantly come up in class discussions, but lacking a solid training in these areas, most physical scientists will simply acknowledge the importance of these fields in dealing with the questions at hand, and then focus the class back on the physical sciences. The same, with the roles of the physical and social sciences reversed, can probably be said of classes in the social sciences. In order for an interdisciplinary education in environmental studies to succeed, it must ultimately blend courses in the physical sciences and the social sciences. How well this can be done depends, in turn, on the ability of the faculty to be interdisciplinary. Given that today's faculty were trained in traditional disciplines, this is more of a challenge than one might think. Perhaps the real success of interdisciplinary programs such as Environmental Studies cannot be measured until a new generation of interdisciplinary thinkers has been produced and takes its place in college faculties. Until then, we must rely upon the vision and enthusiasm of a core group of faculty, and in particular, on the clear enthusiasm of the students, which ultimately drives programs such as Environmental Studies.

APPENDIX I

Environmental Studies Program, University of Colorado at Boulder

The College and Arts and Sciences at the University of Colorado at Boulder offers a Bachelor's degree with a major in environmental studies. Environmental Studies is a program, rather than a department, and utilizes existing course given in various Arts and Sciences Departments. The program is designed to provide a student with (i) a broad but rigorous education in environmental issues and problem solving, as opposed to a traditional, disciplinary-based training and (ii) the disciplinary rigor required to gain employment in the field, but taught in the context of an interdisciplinary education. The environmental studies major is administered through the Office of Interdisciplinary Studies, and draws from curriculum in the earth and natural sciences as well as the social sciences.

The program is composed of a required common curriculum which exposes all students to the basics of physical and social environmental sciences, and a choice between two tracks. The environmental science track has specializations in water, biogeochemistry, and climate, and the social science track has specializations in environment and natural resources, environmental analysis, and decision-making, planning, and policy.

The purpose of the major is to train students in the causes, scale, and remediation strategies of the major environmental problems of the United States and the world. Students will acquire an awareness of the complexity of factors relating to human interaction with the environment, especially the fact that environmental problems have both human and biophysical components, and knowledge of the general principles of human-environmental interaction, global habitability and environmental change, and sustainable human societies.

Students who wish to also pursue a traditional, discipline-based education are encouraged to double major or complete a minor in one of the participating departments. An Internship Program is offered to provide the upper level student with practical experience as well as valuable contacts in the field.

To complement its curriculum, the Environmental Studies program offers its majors internship opportunities, in which students earn academic credit while working in selected positions with private and public agencies and firms. Minimum time required is 10 hours per week for 16 weeks to earn three units of course credit. The primary goal is to provide academically-supervised opportunities for majors to work in public and private organizations on projects related to students' academic interests, in order to relate theory to practice as an expansion of learning and understanding.

All courses used for the major must have C- or better grades. No Independent Study hours apply to the major. No pass/fail hours apply to the major. A minimum grade point average of 2.0 in all C.U. courses is required.

The E.S. major has recently been redesigned to, in part, better equip students to work in the society/policy and/or sciences in the general area of the environment. The program actively tracks its majors after they graduate to access exactly the career opportunities. Students should consider that in some fields an advanced degree may be desirable or even necessary.

Based on past experience with the Environmental Conservation Program students have been employed in the following areas: water resources, waste management, environmental journalism, environmental law, cartography, public planning, computer mapping, market analysis, and handling hazardous materials.

Geographical Concept of Training in Environmental Management: Local, Regional and Global Aspects

Serge G Touchinski
Department of Environmental Management
Faculty of Geography
Moscow State University
119899 Moscow, Russia

One of the most important tasks in global change science is an appropriate description of the environment to enable forecasts to be made of future developments. Environmental management is a very complex and multidisciplinary field of enquiry and consequently the training of experts and scientists who can be the multidisciplinary team leaders in environmental management issues is very important issue.

A multidisciplinary approach in environmental management can have as its focus natural or social sciences. However, combining both foci could create a particularly successful route. One way to do this is to use a geographical approach, because by this way we would have the real combination between nature and society at local and regional levels. For global issues, we can focus our knowledge on earth sciences.

There are different approaches that can be applied in research and training activities in environmental management:
Technological concepts - based on the technological aspects of public production and its ecological basis, such as non-waste or low-waste technology and closed-cycle technology. This can be called industrial ecology.
Biological concepts - concerned with and focussed on ecological aspects such as biocenotic and biospheric interactions.
Socioeconomic concepts - based on the analysis of the economic impacts on the biosphere, including human society, economic structure and development.
Human-health concepts - constructed on health protection and preservation of the society from waste, industrial and other impacts.

The geographical concept of environmental management starts from assumptions of complex interactions between natural phenomena without or

NATO ASI Series, Vol. I 29
Global Environmental Change in Science:
Education and Training
Edited by David Waddington

with socioeconomic impacts. In contrast to concepts described above, the geographical approach based on earth sciences is concerned with natural variability under anthropogenic impacts, taking in account all aspects of environmental management.

A curriculum in Environmental Management with emphasis in global change problems is being developed here in two main directions. One is in training leaders of multidisciplinary teams to tackle global change problems. The second is in training the specialists for such teams.

We have included, for background knowledge, a strong education in the core of earth science disciplines with a concern for socioeconomic disciplines. So the main focus is on the natural sciences (mathematics, physics with elements of geophysics, chemistry with elements of geochemistry, biology, general ecology) and core earth sciences (geology, geomorphology, meteorology and climatology, hydrology, pedology and soil geography, etc.). All these courses are traditionally in courses on physical geography and in geophysics (for global problems) and some parts are in civil and environmental engineering courses (Appendix I).

This part of traditional education is supplemented by a very important and really core discipline, Environmental Basis of Earth Sciences - Developing and Changing Earth. This course is being constructed as selected units describing the most important processes that are studied in the earth sciences, time and space scales of influences of anthropogenic impacts based on the core projects that are included in the International Geosphere Biosphere Program (IGBP 1992).

An introductory Environmental Management course gives the current views of the problem and another course, the History of Environmental Management, gives a background of the interaction between civilization and the environment. The course in the History of Material Culture helps us to create the understanding of long-term structure in the global change processes. Core socioeconomic courses include environmental basis of production, environmental economics, environmental law, social psychology. At the final stage of training, there will be some seminars in theoretical and current problems of Environmental Management.

An extremely important part of an expert's work is the contact with human society to increase the knowledge in fields concerned with environmental problems, and we are planning to organize training in environmental information including elements of journalism, social psychology, environmental law and the new problems in environmental changes.

IGBP (1992) Climate Change 1992. Intergovernmental panel on climate change. Cambridge University Press

Appendix I

Curriculum in Environmental Management

Titles of Disciplines: courses and seminars	years of education 1	2	3	4	5
Nature Sciences	#				
Earth Sciences	#				
Environmental Management (EM):					
Introductory		#			
Developing and changing Earth			#		
Basic ecology			#		
History of EM		#			
Nature resources management				#	
Seminar on theoretical problems in EM					#
Environmental Basis of Public Production		#			
Environmental Law			#		
Environmental Economy				#	
Environmental Geochemistry		#			
Methods of Environmetal Survey:		#	#		
Environmental monitoring		#			
GIS		#			
Methods of field survey		#			
Analitical methods		#	#		
Data system analysis			#		
Environmental modelling			#	#	
Environmental expertize (Ecogeographical)				#	
Basis of Environmental Management					#
Special Topics and Seminars:					
Engineering geography				#	
Social geography and urbanization				#	
Nature preserves				#	
Human health and the environment					#
Environmentally sound technology				#	
Current problems of EM					#
Regional planning				#	
Environmental information					#

Environmental Monitoring: Special Course for Training Global Change Scientists

Serge G Touchinski
Department of Environmental Management
Faculty of Geography
Moscow State University
119899 Moscow, Russia

Measurements in environmental systems are complex, largely because of uncertainties in spatial and temporal predictions of environmental behavior. The environment is often discussed in terms of single samples or measurements taken in space or time but these results do not give a reliable view of environmental systems. Furthermore the existence of pollution sources may be responsible for nonlinearity in the dynamics of the environment on the background of the natural processes. It is very important to have a wide and adequate data-base of environmental measurements in different natural and man-made situations.

The aim of the Masters course (which will also be used for retraining) is to introduce the main concepts in the organization of monitoring networks.

The course includes 60 hours of lectures, seminars and laboratory work. There is also a 4-week period for fieldwork. Monitoring is the means for the observation, assessment and forecasting of the state of the environment, which can be applied to the management involved in sustainable development issues. The main principles are shown in fig. 1. Among the concepts in monitoring classification, the most convenient is that based on the scaling approach. Global monitoring deals with the background processes for large time and space scales. Ecological monitoring can be described as a matrix involving the scaling of processes or phenomena whether they are local, regional or global.

Scales of variability

One of the most difficult tasks is the optimization of observation networks, including spatial and temporal frequency and accuracy of measurements with actual variability of behavior of the ecosystem. However, there are different scales of fluctuations in environmental parameters, which are described

NATO ASI Series, Vol. I 29
Global Environmental Change in Science:
Education and Training
Edited by David Waddington

graphically in figure 2. In the figure, several types of fluctuation are shown which can be regarded to be on global, regional and local scales.

Experimental evidence is given in the lecture course with research findings in aquatic systems such as river, reservoirs, and lakes and in terrestrial ecosystems. The results help us to design systems of monitoring and observations with the necessary spatial and temporal scales, including a set of defined parameters, sampling sites and time selection, frequency and duration. It is possible to discuss the Complex Monitoring Program (COMMON) for predicting changes in ecosystems at different time and space scales.

Methods of observations

The main objectives for global background monitoring observations are given in the Table below. The specific features of background monitoring are the

- determination of very small concentrations of all parameters (tendency of trends) for a large territory with long time fluctuations;
- influences of pollution sources producing very strong anthropogenic noise in all of the measurements;
- periodicity and accuracy of measurements correlated with fluctuations and sensitivity of the methods applied.

Three conditions must be considered:

- the possible long distance from any pollution sources;
- the observation station must be in a typical biogeographical region.
- this territory must have a good history of research and observations.

Monitoring networks parameters

Parameter of network	Monitoring scales		
	local	regional	global
territory area, km^2	10 - 100	20 - 2 x 10^6	up to 10^7 - 10^8
discreteness of observation, km	0:01 - 10	10 - 500	up to 3000 - 5000
periodicity of processes	days-hours	month-years	up to centuries
frequency of observations	minutes-hours	months-decade	years
number of parameters	10 - 100	100 - 1000	more than 1000
accuracy (of limits)	units	% units	0.1 - 0:01% units

It is very important to consider the problem of choice for sampling in the field and accuracy of measurements (fig. 3).

In the background monitoring of global change, the most suitable methods for observations are bioindication for time series and remote sensing methods for spatial pattern of variability of the territory. In this part of the course, we have seminars concerned with the biomonitoring of large scale variations in the biosphere and in remote sensing.

Monitoring data analysis

Assessment models increase our understanding of the environment and to support decision making. However, there is a need to understand better the interplay between the capabilities of the model and the types of questions that the user or regulator needs to answer. The predictive nature of models is largely based on the choice of model structure, time and spatial scales, process description, and data requirements.

Statistics are an essential tool in all aspects of earth science: time series analysis is needed for the examination of variability or fluctuations structure including trend analysis, coherency and shift of phase in natural and in anthropogenic processes. Multivariate and other advanced statistical methods are often required to disentagnel effects in complex issues; and most environmental topics require that experimental results can be analyzed statistically:

- experimental design and analysis for field studies;
- multivariate methods and time-series analysis methods appropriate for monitoring;
- choosing the most informative parameters, describing the variability of ecosystems under anthropogenic impact;
- adaptation of optimum techniques and appropriate methods of measurements in ecosystems research;
- estimates of the state of the ecosystems for applied ecological studies:
- ecological state forecasting using methods of time series analysis for scenarios of possible natural and anthropogenic impacts;
- ecological standardization for geographical regions;
- ecological assessment;
- ecological zoning;
- ecological mapping.

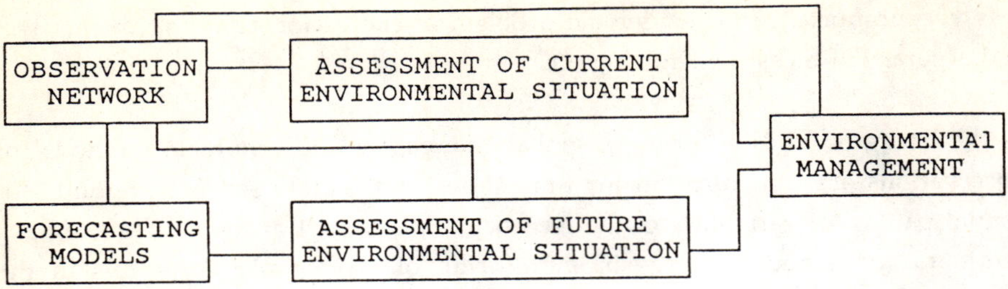

Fig.1 Conceptual scheme of environmental monitoring

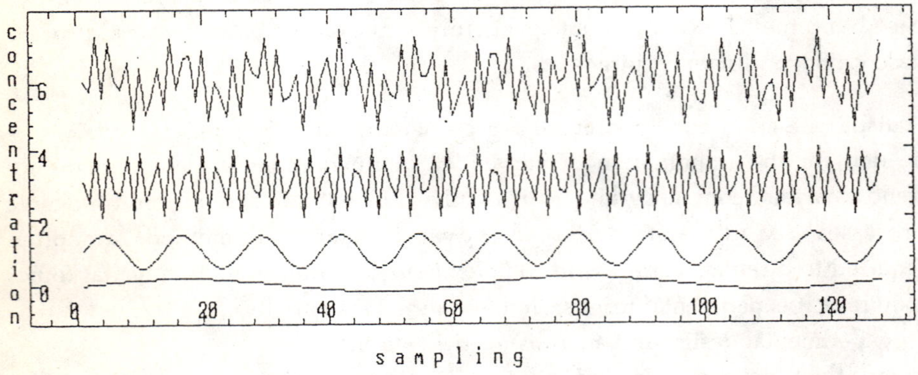

Fig.2 Examples of observations: local, regional and global background monitoring

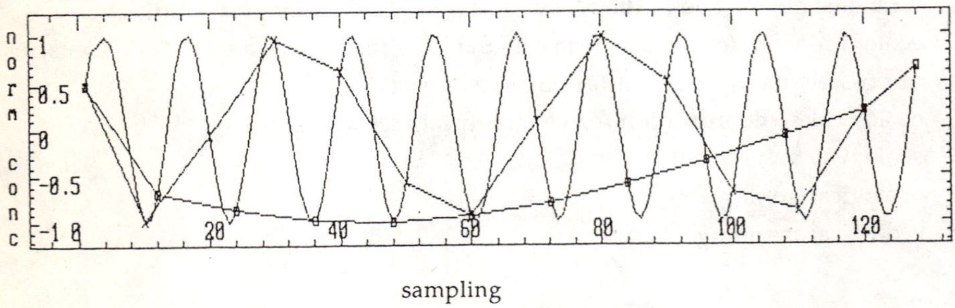

Fig. 3 One source of observation mistakes by using uncorrected sampling frequency

The development of a new masters level course: MSc Environmental Change and Management, University of Oxford, UK

John Boardman
Environmental Change Unit
University of Oxford
Mansfield Road
Oxford OX1 3TB, UK

Introduction

The Environmental Change Unit (ECU) at the University of Oxford is an interdisciplinary unit dedicated to collaborative research on the nature, causes, and impacts of environmental change, and to the development of management strategies for dealing with future changes. The Unit acts as a focus for cross-departmental environmental initiatives within the university. A primary example of this is the development of a new MSc degree.

The Unit was established in 1991 with a management committee drawn from the School of Geography, Department of Zoology and the Oxford Forestry Institute. Research within the Unit is structured within three programmes: long-term environmental change, energy and the environment, and climate impact on crops. It is planned to expand current work in two areas: health and the environment, and land degradation.

Planning the Course

The theme for the degree was relatively quickly arrived at based on the overall remit of the Unit, the expertise available in Oxford and the nature of existing degrees in the UK. It was agreed that "environmental change" would be broadly defined, that the course would concentrate on the present and the future, and that it would encompass both human and physical dimensions. The rejection of "past environmental change" as a theme was mainly because other degrees and training programmes cover this area. The concentration on the present would also be relevant to the management theme of the degree.

From the outset it was argued that if the degree were to succeed, it should aim to attract students who wished to be involved in the management of changing environments at a planning and policy level rather than at a technical level. The ideal student would be one with a strong disciplinary background in earth

NATO ASI Series, Vol. I 29
Global Environmental Change in Science:
Education and Training
Edited by David Waddington

sciences, economics, or ecology, for example, and who wished to broaden his or her experience. At the same time it was hoped that the degree would attract applicants with work experience in environmental management and who wished to update or gain a more theoretical background. The broadening experience was felt to be important: those with strong science backgrounds would be introduced to economic, legal, and policy issues.

The taught elements of the course (lectures, seminars, tutorials, laboratory and field sessions, and visits) occupy 10 to 12 hours per week over three eight-week terms. Courses are examined in the third term and the dissertation is completed in the period from June to September. All elements of the course are designed specifically for the MSc degree and are not part of undergraduate degrees. The initial planned intake is 15 students.

Course Content

The principle underlying the selection of course content was that the degree should contain broad-based core elements that would be undertaken by all students, plus sufficient optional and more specialist elements to satisfy students of very diverse background.

The MSc degree is a one-year course which will run from October to the following September when the dissertation is handed in.

The degree consists of a series of compulsory courses:

Course 1: Issues and Driving Forces

This is a lecture programme which considers the issues and driving forces underlying environmental change. After a general introduction considering central issues, such as sustainability, thresholds, population ecology, and energy, the course is sub-divided into terrestrial, atmospheric, and aquatic areas. These sections include issues such as desertification, salinization, greenhouse gases, (ozone and CFC's), and pollution.

Course 2: Managing the Environment

This is a lecture-based course which is divided into two sections: cross-cutting dimensions, which includes consideration of ethical, legal, and cultural aspects of environmental management; and the institutional framework. The latter deals with international institutions, special interest groups (NGO's, lobbies, and consultancies), and the framework of business and commerce. Finally, the course takes the use of energy as a case study in managing the environment.

Course 3: Methods and Techniques for Environmental Management
This is a lecture, laboratory, and field-based course which begins with general issues, such as experimental design, surveys and monitoring (both of the social and physical environments), and environmental and risk assessment. In the second and third terms, management techniques such as Geographic Information Systems (GIS), remote sensing, and modelling are covered.

The course includes two series of seminars, one by guest speakers which has been running since the founding of the ECU, and the other a student-led series.

A fortnightly series of visits to institutions working in the field of environmental change and management is also part of the course. These include Kew Gardens, National Rivers Authority, Bristol Energy Centre, and IUCN (Cambridge). A weekend field course at Slapton in Devon takes place at the beginning of the course and introduces the students to the management of coasts, nature reserves, and the threat of nitrate pollution to fresh water ecosytems.

Optional elements in the course are important as they allow students to follow their own interests at greater depths. All options develop from topics introduced in the main courses.

An option consists of four tutorials or seminars, depending on the number taking the course. For each option the student produces an essay which forms part of the degree assessment profile. Each student takes two options, preferably one in each of the first two terms. The options offered in 1994-95 are environmental law, conservation and diversity, transport and the environment, environmental education, energy and the environment, drainage basin management, further GIS, soil erosion and conservation, forests, food security and the environment, environmental economics, and cultural adaptation and the environment.

The final element of the course is a 15,000 word dissertation on an aspect of environmental change and management.

Discussion

The MSc Environmental Change and Management does not begin until October 1994, but some tentative conclusions can be drawn from the student selection process. In February-April 1994, 23 applicants were offered places from a total of about 85. Of these, seven are currently completing first degrees, but the majority have some form of work experience. Thirteen are non-British students, with the strongest representation from Australia (3) and the USA (3). Good applications were notably lacking from Africa, Latin America, and Asia, excepting Japan; largely, one suspects, reflecting the lack of educational opportunity and funding. Two scholarships for students from Thailand are available (Riche-Monde Scholarship), together with one rising to two in 1995, for students with an interest in Commerce/Industry and the Environment (Unipart International Scholarship). The bias toward first world students is unfortunate in a course dealing with environmental change, since many of the most urgent problems occur in the third world. The problem will only be addressed by obtaining sponsorship for high quality students from specific disadvantaged areas.

The application form asked applicants to give reasons why they wished to take the course. About 30 applicants were interviewed. It was revealed that there was a strong desire to undertake a broad-based course which also offered opportunities for specialisation. Breadth and interdisciplinarity were seen as positive reasons for applying. Many applicants commented favourably on the inclusion of such areas as economics and law; others welcomed the opportunity to become acquainted with techniques such as GIS and remote sensing.

The content of the course is strongly influenced by the requirement to spread the teaching and supervision load as widely as possible. This is because most of the teaching will be undertaken by staff with existing full-time university appointments. Research staff contribute a minor amount. The reason for this policy is to achieve staff continuity from year to year and to reduce the resource implications that would arise if new staff were to be appointed specifically to teach on the course.

From the course planning point-of-view, the breadth of expertise resulting from resource constraints is not unwelcome in a course of this nature and it also seems to be attractive to prospective students. However, the danger is one of fragmentation, with individual teaching sessions tied together, at best, by broad themes such as critical issues, managing the environment, and methods and techniques. Until students confront optional courses and their dissertation, there is a possibility that the educational experience will lack coherence.

Course planning produced several surprises. Within university departments a breadth of expertise concerning the environment was revealed. However, there were difficulties in filling some areas of the syllabus, such as environmental economics and law. What was also surprising was that there was far more expertise in the area of environmental change than in management issues. Apparently most academics find it easier to study change than to discuss how the environment can and should be managed!

Conclusion

The opportunity to develop a new masters course in Environmental Change and Management alongside the establishment of a new Unit at the University of Oxford is an exciting educational challenge. Not only does it allow the Unit to become a fully integrated teaching part of the university but requires it to integrate expertise in many areas of environmental teaching and research around a theme. The development process has not been without its problems, particularly in the area of producing a coherent course and obtaining sufficient resources to support the course. However, the admissions procedure has already shown the great demand for such a course and the high quality of potential students.

SECTION 3: RESOURCES FOR TEACHING GLOBAL CHANGE SCIENCE

Teaching Global Change Science to Non-Science Majors

George W Fisher
Morton K Blaustein Department of Earth and Planetary Sciences
The Johns Hopkins University
Baltimore MD 21218
USA

When the Apollo astronauts brought back the first photos from space more than twenty years ago, they gave us a new vision of Earth as a system. Through a gossamer-thin layer of air we saw a complex biogeochemical system of water, rock, and soil, inhabited by billions of fellow creatures, large and small, all dependent on one another. Those same astronauts showed us just how important that system is to us by bringing back images of the lunar landscape, barren and lifeless.

The Earth photos showed us that one of our most important jobs is to understand how the earth system works and how humans interact with that system. That task has two implications for the education community. We must, of course, train a new generation of scientists who are experts in the traditional earth sciences or in ecology, and who share a global perspective of how their discipline meshes with the rest of the earth system. But if we are to make effective use of the science these students are to produce, we will also need to involve the social sciences, the humanities, and even the arts as well. We need economists to help us understand how economic systems motivate behavior, and how we can use those systems to motivate people to behave responsibly. We need sociologists to help us understand different cultural attitudes toward nature. We need political scientists to teach us the political and diplomatic side of implementing well-intended measures. Artists, poets, and theologians can help people to see the world afresh and to understand the nurturing role that we now sense the need to adopt.

Consequently, we must teach some of the elements of earth system science to non-scientists, specialists who will help us deal with the practical issues of sustaining Earth, and a generation of citizens who must understand what we are about, if they are to pay for the task. We must convey our present understanding of global systems, but we must also impart an appreciation for the uncertainties in our understanding, the techniques by which we attempt to learn about earth

NATO ASI Series, Vol. I 29
Global Environmental Change in Science:
Education and Training
Edited by David Waddington

systems, and the degree of confidence that we have in our conclusions. In other words, we must communicate an accurate sense of the methods of science and a critical understanding of how effective those methods are, and where they can go wrong.

Too many introductory science courses focus on conveying current interpretations of a discipline as it is now understood. It is far harder, but essential, to convey a feeling for the culture of science, and for how science works.

The method of science - the interactive posing, testing, and refining of hypotheses or models - is easy enough to communicate in principle. But for students to really understand how the method works, and certainly for them to understand how models link to the natural world, they must themselves participate in building and refining models. In the sciences relevant to global change, working with models poses a problem because even the simplest models require levels of mathematical understanding that are typically well beyond those of most humanities and social science majors.

The new graphical differential equation solvers like Simulink[1] and Stella[2] provide an ideal vehicle for overcoming the mathematical difficulties involved in creating and refining models, and permit students to become personally involved in building and testing them. In the process, students can acquire a feel for how models (and nature) work which far exceeds that conveyed by more conventional methods. Few (1991) and other participants in the Earth System Science Education program developed by the Universities Space Research Association (Fisher, this volume) have made considerable use of these graphical equation solvers in laboratory exercises. But the point that I wish to make here is that these same programs are also excellent for communicating the methods of model development to students in large lecture courses that lack laboratory sessions. A much broader range of students can therefore be addressed. With the aid of equipment capable of projecting a computer display on a classroom screen, the instructor can show students the graphical representation of a model, explain the logic of the model, and generate graphical output intended to

---

1 Simulink is a trademark of The MathWorks, Inc., Cochituate Place, 24 Prime Park Way, Natick, MA 01760.

2 Stella is a trademark of High Performance Systems, Inc., 45 Lyme Road, Hanover NH 03755.

represent some natural process by running the model. If the results do not fit natural observations well, the instructor can then engage the students in a dialog to determine how to improve the model: by adding additional elements, by modifying parameter values, or both. By participating in that kind of a discussion, proposing changes in a model or in model parameters, then re-running the model to see precisely what the effects of those changes are, students can develop a real sense of how science is done and how the natural process being modeled actually works.

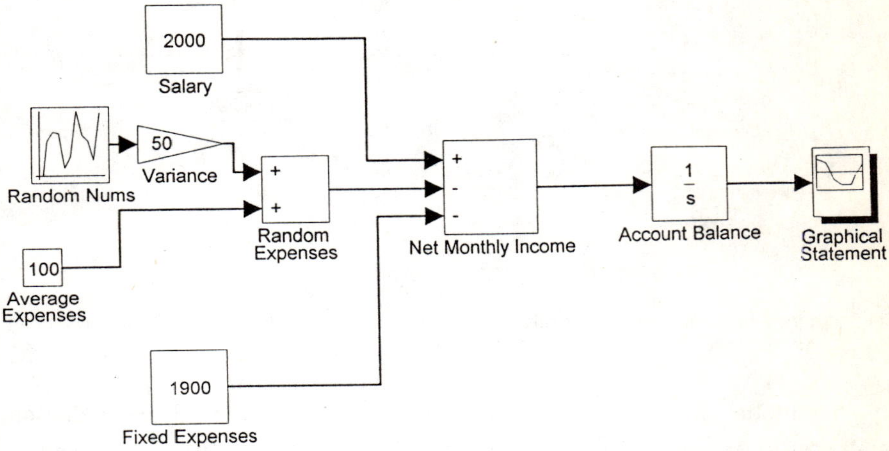

Figure 1: Simulink model of bank balance subject to random impulse buying, used to illustrate concepts of system modeling

To introduce the method to students, I generally use a simple model of a bank account that produces a graph of the account balance over time (Figure 1). The model has most of the elements needed to represent a biogeochemical cycle. There are sources (income), sinks (fixed expenses), parameters which control the size and variability of random expenses, a reservoir (an integrator which determines the current balance in the account), and a graph generator which shows the balance over time. Viewed in this context, the concept of integration seems natural to anyone who has ever balanced a checkbook, but to keep things friendly, I refer to the integrator as an 'accumulator', rather than an integrator. By tweaking the average value and variance of the random expenses and watching how the trend of the monthly balance changes, I can give students a feel for the dynamics of the system, and a sense of how to work with models of natural systems, all in a familiar and comfortable context.

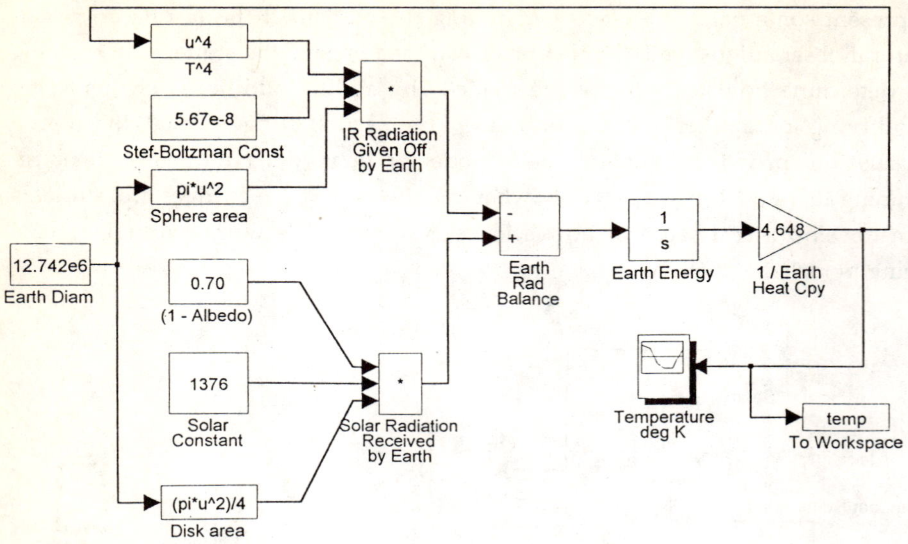

Figure 2: Simple radiation model for calculating Earth's surface temperature in the absence of an atmosphere

From that model, it's a short step to a simple model of the Earth's radiation balance without an atmosphere (Figure 2), and from there to a more sophisticated model that incorporates the effects of an atmosphere (Figure 3). After using the latter model to explain the role of the atmosphere in warming the Earth, we can go farther afield, changing the parameters to simulate conditions on Mars and Venus, and (by reducing the value of the solar constant) document the need for additional atmospheric warming in the early Precambrian.

I've developed Simulink models for processes as complex as ENSO (based on the model of Battisti and Hirst, 1989), and have found students willing to go a surprising distance into some of the more arcane realms of modeling when given a tool that instantly produces visual results to which they can relate. In the process, they seem to gain a real understanding of the modeling process, of how we can use models to refine our interpretation of complex natural processes, and of how the processes themselves work and respond to perturbations. I am convinced that this approach gives a better understanding of how Earth systems work and how we study Earth systems, than any other method with which I am familiar.

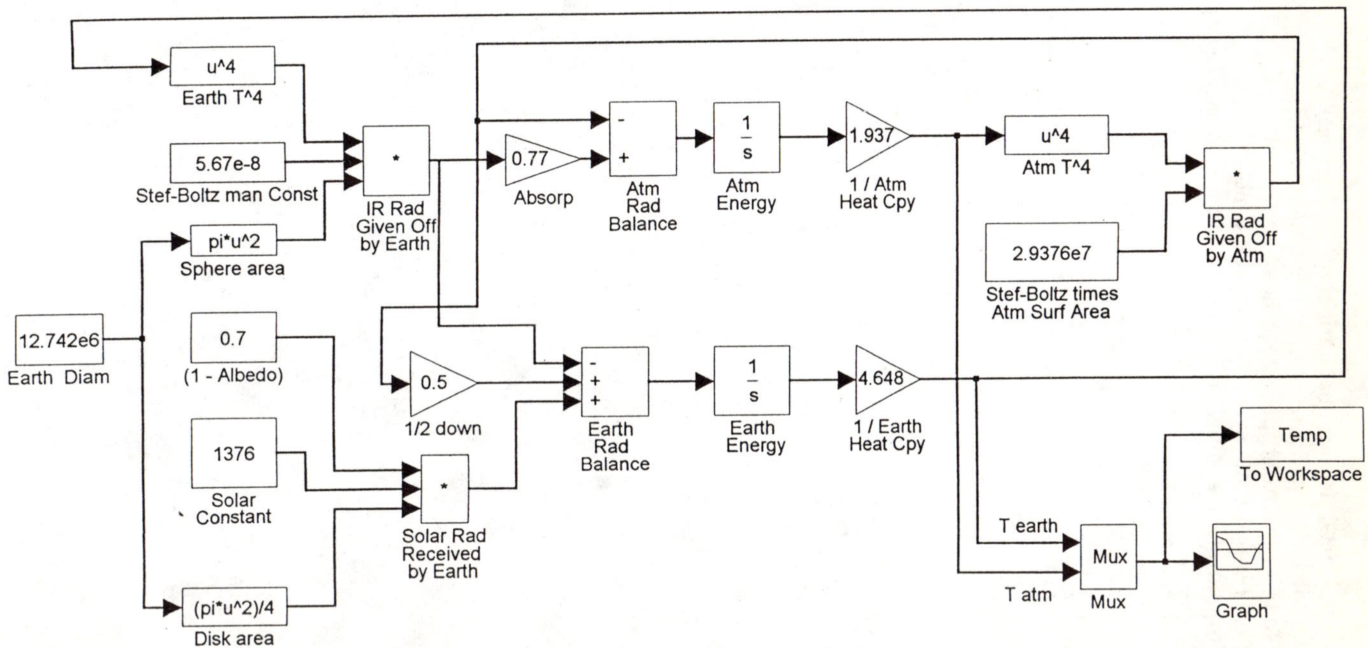

Figure 3: Radiation model for estimating Earth's surface temperature in the presence of an atmosphere

References

Battisti DS and Hirst AC (1989) Interannual variability in a tropical atmosphere-ocean model: influence of the basic state, ocean geometry and nonlinearity, Jour. of the Atmospheric Sciences 46: 1687-1712

Few AA (1991) System behavior and system modeling. Global Change Instruction Program, University Corporation for Atmospheric Research, Boulder CO

Teaching the Human Aspects of Earth System Science

Catherine Gautier, Diane Schweizer and William O'Hirok
Earth-Space Research Group
CRSEO - Ellison Hall
University of California
Santa Barbara, CA 93106, USA

One of the issues facing educators of Earth System Science is how to teach the human aspects of this discipline to students having mostly science backgrounds and interests. To address this issue, a "mini Rio Summit" is now an integral part of the Earth System Science curriculum at the University of California, Santa Barbara. The mini Rio Summit offers an avenue for students to investigate issues of development with limited environmental degradation and to confront some of the intricacies of international environmental decision making. The summit encourages dialogue among students, who represent nations participating in this summit, on topics of global change. It also broadens their awareness of political and socioeconomic considerations that individual countries must face when addressing global change issues. Students' evaluations of two summits (1993 and 1994) were positive and enthusiastic. Their response indicates that the summit is a successful method for bringing them more in touch with the complexities involved in dealing with both the physical and human aspects of global change.

Introduction

New Earth System Science (ESS) classes including Earth Sciences are emerging as part of the curriculum in many departments, and educators are faced with how to teach the human component of this discipline to students who have mostly science backgrounds and interests. To demonstrate the complexities associated with international decision making in the face of global environmental change, a "mini Rio Earth Summit" has been integrated into the ESS course recently developed in the Geography Department of the University of California, Santa Barbara (UCSB). Just as the Rio Earth Summit held in Rio de Janeiro, Brazil in June 1992, provided a forum for the nations of the world to voice and share their opinions on environmental protection and the move towards sustainable development, the UCSB mini summit provides a means for students to develop an awareness of concerns about global environmental change and to confront some of the intricacies of international decision making.

NATO ASI Series, Vol. I 29
Global Environmental Change in Science:
Education and Training
Edited by David Waddington

Earth System Science at UCSB is a recently developed (1992) course which provides the foundation to study the earth as a system and describes how it has evolved and continues to evolve under the constraints of nature and human activities. ESS presents the earth's atmospheric system as an integrated system of physical, biogeochemical, and human components, rather than as a group of independent subsystems. As with many such courses, the immediate question is how to incorporate human, social, and economic aspects of the evolving world in a curriculum which mostly concentrates on the physical basis, the biogeochemical aspects, and the linkages between components of the earth system. The mini Rio Summit concept has been developed in response to this need.

The Mini Rio Summit

a. Summit objectives and organization

The objective of the mini Rio Summit are: (1) to prepare students to be informed citizens in a global world, and (2) to enhance student awareness of difficulties associated with decision making on an international scale. The mini Rio Summit constitutes the final exam for ESS. Students are informed of the project on the first day of class. Throughout the quarter, class discussions are held concerning the content of the summit. One month prior to the initial meeting, students select the specific nations to be represented, with guidance from the instructor only to ensure social, demographic, and economic representation.

The mini summit is organized so that student groups represent particular nations or geographical regions which are economically, environmentally, and politically dissimilar. For example, Japan, China, Tanzania, Russia, Mexico, and New Zealand were the countries chosen by students to be represented at the 1994 mini summit. Students are required to become familiar with the environmental, economic, social, and political status of "their" selected country, as well as that of the other countries at the summit, so that they can debate an individual country's path towards sustainable development. *World Resources: A Guide to the Global Environment* (Hammond 1992 and 1994) and *Confronting Climate Change* (Miller 1992), are used as the main resources for preparation of the mini summit. These texts contain a collection of in-depth articles and environmental facts.

Students participate in the mini summit by putting themselves in the position of the nation they represent and by offering their opinions as to how topics of environmental concern could be handled, given the nation's current social, economic, and political circumstances. Each nation is required to think in terms of how it could move towards its own version of domestic sustainable development within the global community.

The mini summit stresses the importance of global interactions and communications. Students are informed of the positions of the other countries to better understand the paths of action of other countries and the consequences of those actions. As part of a global community, representatives are encouraged to call upon other nations for support and to offer their support to other nations, as well as to reproach them for unsound environmental practices.

There are two meetings of the representatives. The initial meeting is a "statement of position" in which students present an overview of each country's current environmental and economic circumstances. The second meeting takes the form of a panel discussion, which is designed to allow more in-depth discussion of selected topics and interaction among representatives. This second meeting is designed to be a precursor to a larger global environmental meeting.

b. Presentation of Environmental Position

During the first meeting, representatives deliver their country's position on topics which are of environmental significance to them. Topics include, but are not limited to, population, health, biodiversity, pollution, food, energy, and freshwater availability. The presentations include the political, economic, historical, and social circumstances facing the countries and are supported by recent statistics. The students are asked to offer specific reasons as to how the current situation has come about.

The position of each nation is presented to the representatives of the other countries as well as to additional members of an audience that consists of graduate students in the Geography Department. After the position of each nation is delivered, members of the audience are given the opportunity to initiate discussion and ask detailed questions concerning the contents of each presentation.

The presentation of the environmental positions increases student awareness of the complexities involved in developing environmental policy on an international scale. Students are able to identify particular areas of concern, present strategies for dealing with these concerns, and confront any corresponding policy problems. Successful steps that were decided upon and enacted in order to address the issues of concern are presented and discussed. Students become more familiar with the conditions of the other nations, which helps them formulate more encompassing opinions.

This first meeting of the mini summit provides a solid foundation for the second meeting, the panel discussion. The presentations enable individual countries to become familiar with many facets of the same issue. The panel discussion is held one or two days later, allowing students time to digest the information received and to prepare appropriate questions and concerns to be addressed.

c. Panel Discussion

The purpose of the panel discussion is to encourage communication among the individual nations, to outline the concerns of the global community, and to prepare for the necessary bargaining that typically takes place in establishing policies and treaties. During the presentation of environmental concerns, students usually develop a nationalistic attitude and become interested in exploring global environmental change from 'their' country's perspective, as well as from a global perspective.

The panel discussions are led by a moderator and include the representatives of each nation and an audience of geography graduate students. A week in advance, students are given a list of potential topics/questions to be addressed in the panel discussions. These questions are provided to help students prepare for the panel discussion and to give the discussion focus. The topics of the panel for the 1994 summit were:

1) cross-border pollution and water resource utilization
2) role of multi-national corporations and the environment
3) equality of energy resource use
4) saving the environment for future generations.

A couple of days prior to the discussions, several specific questions are provided to fine-tune the summit presentations.

The day of the panel discussion, the moderator initiates discussion by asking the panel one of the questions. Each 'nation' is given two minutes to respond. After all responses are heard, the floor is opened up for discussion and comments by, first, the representatives of other 'nations' and then the members of the audience.

The panel discussion delves into the intricacies of achieving sustainable development and leads to heated dialogue as to what are the optimal paths for each country to follow.

At the end of the 1993 panel discussions, each nation's representatives were asked to offer at least one plan of action they thought to be the most realistic for minimizing detrimental global environmental change. They were also asked to discuss the limitations of the plan and requirements for putting the plan into action. Some of these action plans were:

1) Fund research for alternate energy sources
2) Cease environmentally unsound fishing practices
3) Develop higher quality mass transit systems
4) Implement mandatory recycling programs
5) Encourage market mechanisms for pollution control
6) Switch from logging to harvesting forest products
7) Promote public education on achieving sustainable development

Overall, a panel discussion provides an excellent forum for students to express their opinions and concerns on global environmental change. If they do not agree with the policy of a particular nation, they are given an opportunity to voice their concern and to offer alternative suggestions. The interaction of many different viewpoints forces students to think carefully about their potential actions and how other countries are likely to respond to those actions.

Discussion

The response to both the 1993 and 1994 mini summit experiences was overwhelmingly positive and enthusiastic. Students felt the inclusion of the mini summit in a science course helped bring out the realization that the earth is both a physical and a human system, and that the two components are not as separate as they might first appear. Students indicated that they became more aware of the problems that can arise when trying to transcend national boundaries and work for the benefit of the international community. The mini summit works well because students generally develop a strong nationalistic attitude towards the nation they represent. Students are quick to learn how difficult it can be to put nationalistic attitudes aside in order to allow an international spirit to reach full bloom.

During the 1993 mini summit, all of the students' viewpoints were "green", indicating a tremendous desire on their part to limit environmental impact and strive for sustainable development. For the 1994 mini summit, a number of students were asked to introduce more controversy into the summit discussions by voluntarily expressing "non-green" attitudes. Ways to include economic and business students in future mini summits will be investigated.

Besides heightening student awareness of enacting public policy and educating them other nations' viewpoints, participation in the mini summit results in several fringe benefits. These benefits include sharper dialogue skills and an understanding of the importance of being well prepared for public debates. The summits were particularly successful as part of the upper division ESS course because the students were genuinely interested in the topics of discussion and devoted considerable time preparing their position papers, as well as their questions/concerns to discuss with other 'countries'. In general, the opinions expressed were all extremely encouraging and suggested that more courses should include this kind of exercise.

References

Hammond AL (ed) (1992) World Resources 1992-1993: A Guide to the Global Environment, Toward Sustainable Development. The World Resources Institute in collaboration with UNEP and UNDP. Oxford University Press, New York

Hammond AL (ed) (1994) World Resources 1994-1995: A Guide to the Global Environment, People and the Environment: Resource Consumption, Population Growth, Women. The World Resources Institute in collaboration with UNEP and UNDP. Oxford University Press, New York

Miller IM (ed) (1992) Confronting Climate Change: Risks, Implications and Responses. Cambridge University Press, Cambridge

Historical Geology: An Avenue for Teaching Global Change Issues

Nancy Healy-Williams
Marine Science Program*
University of South Carolina
Columbia, SC 29208 USA

Introduction

For several years a traditional historical geology course has been offered to non-science majors enrolled in the South Carolina Honors College at the University of South Carolina. During the spring 1992 academic semester that course was revised to reflect the growing trend of examining the earth as a system of interrelated components.

Global change is viewed by many scientists as representing a critical juncture in the history of the earth which presents numerous challenges to the world's populations. Historical Geology appeared to be a perfect avenue with which to explore earth system science in terms of past and future global change processes. The topics covered in the course by no means offer an in-depth examination of the issues of global change but do raise the level of understanding of the issues for non-science majors. The response to the course has been overwhelmingly positive.

Historical Geology was designed to reach several goals. First, students were to learn that earth has a history that is not just geological, but also multifaceted. The earth was to be examined as a system of interrelated components: biosphere, geosphere, hydrosphere, atmosphere, cryosphere, and the biogeochemical cycles. Second, it was hoped that students taking the course would attain a knowledge base from which they could make informed decisions regarding global change issues. Scientists have a duty to educate the general public to help them become more scientifically literate, particularly with the important issues of global change. This historical geology course represents one approach by which this goal can begin to be attained. What follows below is a brief discussion of the topics examined and the teaching methodologies employed.

---

* Present address: South Carolina Commission on Higher Education 1333 Main Street Suite 200 Columbia SC 29021 USA

NATO ASI Series, Vol. I 29
Global Environmental Change in Science:
Education and Training
Edited by David Waddington

Teaching Methodologies

The format consists of lectures, cooperative learning groups, outside reading assignments, class discussions, videos, and exercises drawn from "Education for Global Change". By utilizing several teaching strategies it was hoped that the students would have a keen level of interest in the subject matter. Listed in Table 1 is the topic schedule and reading assignments for a twice weekly class meeting. The textbook used is *Exploring Earth and Life Through Time* (Stanley, 1993).

The course begins by developing an understanding of the physical characteristics of the earth. For example, during the "Dynamic Nature of the Earth" topic students are introduced to the theory of plate tectonics and also a discussion of how altered continental configurations can affect climate patterns. Next we examine the materials of the earth and then delve more deeply into how these materials are used to decipher earth history in both qualitative and quantitative terms. Once a level of understanding is developed concerning the geosphere, these non-science majors (ranging from freshman to seniors) explore how the earth's systems are connected and how feedback mechanisms play an important role in the evolution of those systems. During this portion of the course, exercises are incorporated from "Education for Global Change", which allows a break from the lecture style delivery and also provides a format for class discussion (Stoltman and Waddington, 1994).

Table 1
Historical Geology

Introduction to the course

Lecture Topic Schedule*	**Reading Assignments**
Characteristics of the Earth	Chapter 1 (p. 1-3; 13-16)
Dynamic Nature of the Earth	Chapter 6, Chapter 7 (p. 157-167)
Materials of the Earth	Chapter 1 (p.3-6), Appendix 1
Sedimentary Environments	Chapter 3
Time and Geology	Chapter 1 (p.16-18), Chapter 4
Evolution and the Fossil Record	Chapter 1 (p9-12), Chapter 5
Introduction to Earth System Science	Outside reading
The Earth System: Global Atmospheric Processes	Outside reading
The Earth System: Global Ocean Processes	Outside reading
The Earth System: Global Hydrologic Cycle	Outside reading
The Earth System: Global Carbon Cycle	Outside reading
The Earth System: Paleoclimatic Change Processes	Outside reading
The Mesozoic Biosphere-Atmosphere System	Chapters 13 and 14
The Cenozoic Biosphere-Atmosphere System	Chapters 15 and 16
Final Exam	Modern Global Change Issues and Challenges

* Some topics are two class periods

The earth system science sequence is concluded by examining the tools and methodologies used by paleoclimatologists, paleoceanographers, and paleoecologists. Students are told that paleoenvironmental investigations are like solving a puzzle or mystery and to successfully do such investigations they must become great detectives. Consequently, students are frequently forced to rely on information they acquired during the physical geology sequence in order to answer questions posed by the instructor. A typical question asked is how scientists could possibly determine that there was a great interior seaway on North America during the Jurassic period. This type of question allows students to utilize and synthesize materials (presented earlier) into a cohesive framework.

Two classes are devoted to examining the evidence for present day global change. Topics included within this range from global warming to ozone depletion. Students view the video "A Hole in the Sky" (Nova Series) and enter into a discussion of the evidence presented. Unlike typical historical geology courses earth history is examined for the last 245 MA and approached in terms of the climate system. A teaching methodology utilized is cooperative learning exercises based on the Jigsaw Classroom (Aronson 1978) and adapted from a Geological Society of America workshop "Alternative Pedagogies in Geological Sciences". The Jigsaw Activity is described in Table 2 and represents the questions asked during the Mesozoic topic. The instructor interacts with the groups offering guidance when requested. One might ask if such exercises are successful and how well they are received by the students. The responses from my class were:

> "Do more jigsaw activities";
> "I learned more by doing the jigsaws than from listening to a lecture (no offence)";
> "The jigsaws made class enjoyable and the subject easier to learn".

Four such exercises are spread out over the lecture topics and from the student comments these activities are well worth the effort.

Table 2
Historical Geology Jigsaw Activity *

On the attached sheet are 4 questions about the Mesozoic to be completed as homework. Write a short answer (half page max.) to each question on a separate sheet of paper. Bring these answers to class on April 14. At that time we will perform the Jigsaw Activity outlined below.

1. At the beginning of the class period you will meet in your assigned home team. You and your teammates will each be assigned a different question to become an "expert" on.
2. Each expert will then join a group made up of "experts" on the same question. In your "expert" group discuss your answers to your assigned question and formulate a composite group answer. Plan ways to explain the answer to your home team.
3. Return to your home team. Explain the answer to your assigned question. Listen to your teammates' answers to their expert questions. Discuss the answers as a group.
4. Class discussion. I will respond to any questions from the class about the homework.
5. At the end of class, hand in the three questions on which you were not an "expert". You will receive credit of 10 points for every *genuine attempt* to answer the questions
6. Your "expert" question will be re-written as a new and improved answer, incorporating any ideas you received from your classmates.
7. Hand in the re-written answer to your "expert" question on Thursday, April 19. It will be graded for clarity, completeness, and accuracy on a 50-point scale.

QUESTIONS

1) In terms of animal life, what were the effects of the Late Triassic extinctions? Was this the result of climate change?
2) During the Triassic Period both dinosaurs and mammals (and mammal ancestors) lived together, yet this time period is considered the age of the dinosaurs. Why? What factors influenced this rise?
3) The Cretaceous period witnessed a change in floral patterns. What was this change and what reasons are given for its development?
4) What is some of the evidence for warm global climates during the Early Cretaceous?

* Adapted from Ann Bykerk-Kauffman, Cal. State Univ. Chico

The final exam is conducted during the last two days of the lecture schedule and consists of oral presentations presented by groups of students. The exam is adapted from another exercise in "Education for Global Change" (Stoltman and Waddington, 1994) and is shown in Table 3. Students are presented with the exam well in advance in order to prepare their arguments. Groups are assigned randomly and they are provided time in class to organize. Full preparation is done outside of class.

Table 3

Historical Geology

Final Exam

The Global Warming Threat: Is There a Problem?

This semester you have hopefully learned a great deal about the earth as a system consisting of numerous components. By now you should have an understanding that the earth is a dynamic place which has changed throughout its history. The question facing us now is whether or not the human population on earth is altering the system in ways which will adversely affect our lives and that of many other organisms. Is there a problem? If so what should we do about it, if anything?

Below is a list of four possibilities in answering the global warming question:

- We *take action*; the global warming *threat is real.*
- We *take action*; the global warming *threat is not real.*
- We *take no action*; the global warming *threat is real.*
- We *take no action*; the global warming *threat is not real.*

You will be assigned to a group which will be given the task of presenting the position for one of the four scenarios listed above. On Tuesday, April 26, each group will be allowed approximately 15 minutes to make their presentation regarding their scenario. On Thursday, April 28, the entire class will debate (by group) the four scenarios. It is important to remember that there is no pre-defined "correct" answer to a question of this importance and complexity. The four scenarios really represent the questions and opinions being expressed as the world's societies attempt to make a decision and develop a plan of action. The possible implications of each scenario have important ramifications depending on how real the threat of global warming is. One way to think about the scenarios is in terms of costs - possible costs involved both with the problem (global warming), and possible costs involved with actions taken to address the problem. As used here, costs extend well beyond the quantity of money involved. There can be costs associated with changes in life style and other aspects of the catch-all notion "quality of life". There can be costs involving time and labor, costs associated with losses in personal choice, career options, and so on. Costs do not necessarily have to be associated with taking action; there can be costs involved with non action. Think about how potentially serious the problem is and how likely is its occurrence. You can address your scenarios by examining how effective possible remedies or preventive actions are or will be. How willing do you think the world will be to bear the costs?

GRADING:

Each group will be assigned a letter grade for the effectiveness of their argument. This will include presenting a logical and documented argument for their "cause". Individuals within each group must participate in order to be included in the grade, i.e., the group may not rely on a single spokesperson for the group. A limited number may make the initial presentation and then the others may be involved more heavily in the debate.

IMPORTANT:

Anyone absent from class on either of the two days of this final will receive an automatic grade of F for the final.

Laboratory Exercises:

Historical Geology is a laboratory course which entails meeting weekly for two-hour lab sessions. The syllabus is provided in Table 4 and represents a combination of common historical geology exercises combined with global change exercises. The lab manual referred to in the table is by Frey, et al. (1992). The SimEarth labs are computer exercises designed around the software package of Maxis (SimEarth: The Living Planet). This software package allows students to manipulate earth systems as they evolve throughout geologic time. Working in pairs, students must perform specific tasks and explain the response of the system. This exercise proves very useful in demonstrating the feedback mechanisms of the earth's systems.

Table 4
Laboratory Topic Schedule

Topic	Reading
Stratigraphic Relationships	Chapter 6
Sedimentary Parameters and Sedimentary Rocks	Chapters 1 and 2
Stratigraphic Principles	Chapter 4
Time and Geology	Chapter 5
Introduction to Fossils	Chapter 3
Introduction to Earth System Science	Handout
The Changing Atmosphere	Handout
Paleoecology an Environmental Reconstruction	Chapter 12
SimEarth	(two lab sessions)
Mesozoic Life	Chapter 10
Cenozoic Life	Chapter 11

Conclusion

This course has been taught since 1992 and has been a great success with the students. Their feedback has indicated that they now have a better appreciation of the complexity of the earth system as well as the inter-relationships involved in the system. In terms of the primary goal of the course, it has raised the scientific literacy of the students involved. Many students took this class with some trepidation and the belief that geology would be stale and boring. Utilizing a variety of teaching methodologies and making the subject matter relevant to real world problems resulted in an enthusiastic response from the class. Some students actually indicated on their course evaluations that "they had no idea that geology could be interesting and fun". Taking the time and effort to alter class instruction can have its rewards.

References:

Aronson E (1978) The Jigsaw Classroom. Sage Pub Beverly Hills CA

Bykerk-Kaufman A, Savoy LE, Schneiderman JS (1993) Alternative Pedagogies in Geological Science: A Workshop. Geol Sci Amer Boulder CO

Frey Robert W, Pinet Paul R, Whitney James A (1992) Earth History An Introduction to the Methods of Historical Geology Hunter Textbooks Inc. Winston-Salem NC

Stanley SM (1993) Exploring Earth and Life Through Time. WH Freeman New York

Stoltman JP, Waddington DJ (1994) (eds) Science for understanding tomorrow's world. International Council of Scientific Unions

THE TRUTH ABOUT CATALYTIC CONVERTORS
A workshop activity in the area of Atmospheric Chemistry

C Anastasi
Department of Chemistry
University of York
Heslington
York YO1 5DD, UK

Introduction

A course titled *Chemistry, Resources and the Environment,* which includes many issues within Global Change Science, has been described elsewhere in this book. There are many interesting and distinctive features associated with this course, not least the various methods of teaching used. There are the traditional methods of lectures, seminars and practical work; small group teaching via tutorials are also very effective teaching methods, while site visits and 'expert' guest speakers provide valuable broadening activities. Workshop activities arguably provide the most effective methods of teaching in this course. Here, students participating in a particular course unit come together for an exercise that typically lasts about 4 hours. Small groups (typically 3 or 4 students) are formed so that students can share knowledge and experiences in working through an exercise that addresses some aspect of the course.

The informal nature of the environment and the supportive action of working in a team, with the help of tutors as required, is a very powerful learning vehicle. Students are able to develop the ability to interact effectively in a team working towards a common goal, and to learn valuable communication skills: verbal through presentation of group work; written through formal reports for assessment by the tutors.

What is described in this paper is a workshop which services one of the optional units, *Chemistry in the Atmosphere,* which contributes to the *Chemistry, Resources and Environment* course. The workshop covers three different aspects of Global Change Science: acid rain, photochemical air pollution and the greenhouse effect and, in particular, the impact of catalytic convertors on each phenomenon. It is given in some detail to illustrate the depth of understanding required and a particular strategy, which is helpful in teaching and learning 'open-ended' areas of science such as those associated with Global Change.

NATO ASI Series, Vol. I 29
Global Environmental Change in Science:
Education and Training
Edited by David Waddington

Background to catalytic convertors

Exhaust emissions from vehicles contain gases harmful to the environment including unburnt hydrocarbons (HC), carbon monoxide (CO) and oxides of nitrogen (NO and NO_2, known as NO_X). Catalytic convertors were introduced into the exhaust stream to reduce these emissions and are now a standard item in many new cars. The most common is a three way catalyst with a mixture of platinum metals as the active component. The catalysts causes a reaction between oxygen and NO_X, and between CO and the HCs. Oxides of nitrogen are reduced to harmless nitrogen, and the carbon monoxide and hydrocarbons are oxidised to carbon dioxide and water. When working optimally, catalytic convertors are able to remove 90% of the unburnt HC, CO and NO_X. Lead in petrol 'poisons' the catalyst material and to accommodate these devices has meant the introduction of lead free petrol, thus removing the environmental impact of lead at the same time.

Catalytic convertors have been presented as potentially important devices in alleviating the acid rain and photochemical air pollution, where the gases of interest are NO_X; and HC, CO and NO_X, respectively; there has also been a suggestion that these devices enhance the greenhouse effect because they convert HC and CO to CO_2, the latter an important greenhouse agent. This workshop addresses the impact of catalytic convertors in each of these important environmental issues.

The Impact of Catalytic Convertors on Acid Rain

The acid rain debate is one which has gone on for over two decades. The term (acid rain) is commonly used to describe the deposition of atmospheric pollutants, whose acidity causes damage to forests, soils and water, buildings and materials, and agriculture crops. It is an unsatisfactory term. Some pollutants may be acidic when deposited, others become so only after deposition. Moreover, the pollutants are not necessarily carried to the ground with rain, snow or fog; they may also be active in a dry form as gases or aerosols.

Workshop activity

Sulphur containing compounds and oxides of nitrogen are known to contribute to the acid load in the atmosphere and a ratio of $H_2SO_4/HNO_3 \approx 2$ for has been measured in the UK. Using this value and the information given in Table 1 calculate the percentage reduction in acid loading in the atmosphere by installing catalytic converters on all petrol driven cars in the UK. State the assumptions inherent in the calculation.

Table 1: NO_x emissions by source in the UK in 1990

Emission source	Thousand tonnes	Percentage of total
Domestic	68	3
Commercial/public service	56	2
Power stations	777	28
Refineries	36	1
Other industry	231	8
Railways	30	1
Road transport (petrol)	802	29
Road transport (diesel)	579	21
Civil aircraft	14	1
Shipping	131	5
TOTAL	2729	100

The Impact of Catalytic Convertors on Acid Rain - Tutors notes

The H_2SO_4/HNO_3 ratio has been measured to be 2, i.e. 33% of acid is NO_x-derived.
The table shows petrol driven vehicles contribute 29% of the total NO_x.
If catalytic convertors are fitted to all petrol driven vehicles, the *reduction* in 'acid' loading of the environment is 8.6 %.

Assumptions and other comments:

- HNO_3 and H_2SO_4 are equally effective with respect to environmental impact.
- All NO_2 ends up as HNO_3.
- The convertor operates at optimum conditions at all times.
- There are no cars already fitted with catalytic convertors.
- The H_2SO_4/HNO_3 ratio is only an average and there can be significant variations.

The Impact of Catalytic Convertors on Photochemical Air Pollution

Photochemical Air Pollution (PAP) or Photochemical Smog is a phenomenon resulting from the action of strong sunlight on NO_x (NO and NO_2) and hydrocarbon polluted air, leading to reduced visibility and the formation of ozone (a key smog indicator), and other oxidising agents. The problem was highlighted several decades ago in the Los Angeles basin in the USA. However, it soon became apparent that enhanced ozone concentrations were also prevalent in other North American cities. It is now recognised to be a widespread phenomenon, the symptoms appearing when meteorological conditions (e.g., temperature inversions and anticyclonic weather) or geography, or both, are favourable.

Typical atmospheric measurements of the key components HC, NO_x and O_3 are shown in Figure 1. The primary pollutants HC and NO are high in the morning due to an increase in human activity. As the photochemistry progresses, NO is converted to NO_2, and the latter is photolysed leading to O_3 rising in the afternoon. The O_3 level observed in the afternoon is directly related to the HC and NO_x values in the 6-9 am period.

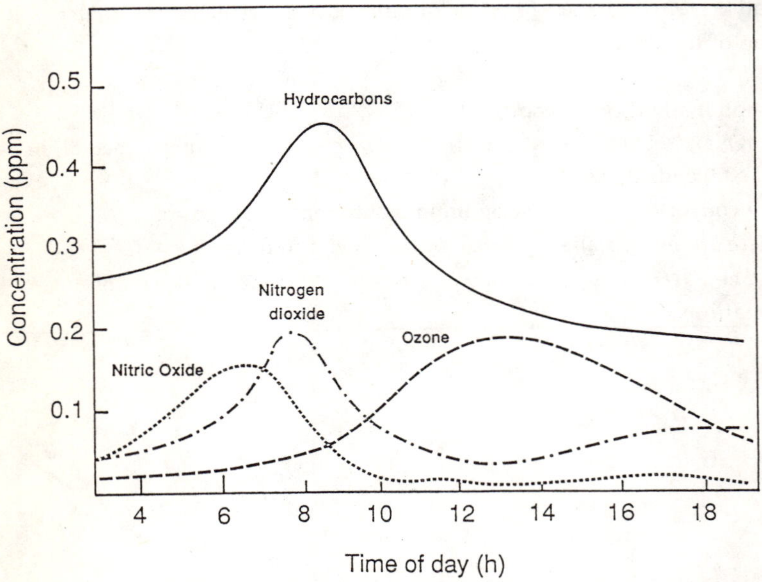

Fig. 1 Ambient measurements of key primary (hydro-carbons, NO and to a lesser extent NO_2) and secondary (O and NO_2) pollutants in an urban environment.

Workshop activity

Figure 2 shows ozone isopleths generated using an atmospheric model. A straight line may be drawn through the apex points of the isopleths dividing the figure into predominantly hydrocarbon-rich and NO_x-rich regions. State what control strategies would lead to a reduction in O_3 levels for these two regions.

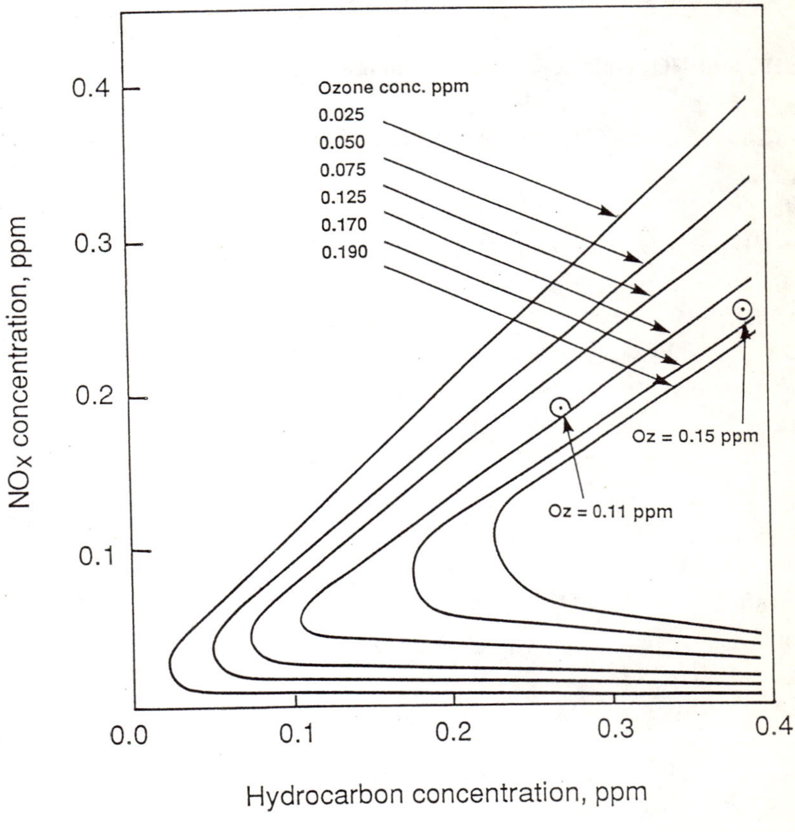

Fig. 2 Ozone isopleth generated using the mechanism in Table 1 under the conditions outlined in the text.

Control strategies for photochemical air pollution - Tutors notes
To the left of the apical axis a reduction of hydrocarbon levels results in the greatest decrease in ozone yields; to the right of this line, a reduction of NO_x levels will lead to the greatest decrease in ozone formation.

As indicated earlier, Figure 1 shows actual HC and NO_x measurements for a typical day in a European city. Calculate the average 6-9 am HC and NO_x values and plot them on the O_3 isopleth, taking note of the O_3 value under these conditions.

Table 2 below summarises the HC and NO_x emissions by source in the UK for 1990. Use this data to establish the reduced values for HC and NO_x emissions if all petrol driven vehicles are fitted with catalytic convertors. Plot these on the isopleth and comment on the 0_3 predicted under this scenario.

Table 2: HC and NO_x emissions by source in the UK in 1990

Emission source	**HC/percentage of total**	**NO_x/percentage of total**
Domestic	2	3
Commercial/public service	-	2
Power stations	-	28
Other industry	2	8
Refineries	-	1
Processes and solvents	50	-
Gas leakage	1	-
Forests	3	-
Railways	-	1
Road transport (petrol)	33	29
Road transport (diesel)	8	21
Civil aircraft	-	1
Shipping	1	5
TOTAL	100	100

The Impact of Catalytic Convertors on Photochemical Air Pollution - Tutors notes
From Figure 1, the mean NO_x ($NO + NO_2$) over the period 6-9 am is Ø.25 ppm and the mean HC over the period 6-9 am is O.39 ppm. From Figure 2, an HC value of O.39 ppm and an NO_x value of O.25 ppm yields an ozone value of ~O.15 ppm (in fair agreement with that observed in the afternoon peak of ~O.18ppm shown in Figure 1).
Using the HC and NO_x values given in Table 2, use of catalytic convertors can reduce the HC from petrol driven vehicles by 0.33 x 0.9 or 30%; HC remaining in the atmosphere is O.27 ppm. Similarly, use of catalytic convertors can reduce NO_x from petrol driven vehicles by 29 x O.9 or 26%; NO_x remaining in the atmosphere is O.19 ppm. From Figure 2, an HC value of O.27 ppm and NO_x value of O.19 ppm yields an ozone value of ~O.11 ppm. Fitting of catalytic convertors on all vehicles would, in this case, reduce the ozone levels to the maximum 1 hour exposure level suggested by the World Health Organisation.
Assumptions and other comments:

- The isopleths given were constructed for Midsummer's Day in an urban environment; the model used detailed chemistry with three representative HC's.
- Different models will yield slightly different isopleth diagrams.
- As the catalytic convertors age they become less efficient and the reduction of O_3 is less.
- In this case, a reduction in HC, with NO_x left alone, would lead to a greater reduction in ozone.

The Impact of Catalytic Convertors on the Greenhouse Effect
The greenhouse effect is strongly linked to the concentration of CO_2 in the atmosphere and, in particular, its strong absorption of long wave radiation emitted by the earth. If sufficient CO_2 is released into the atmosphere by human activities, the radiation balance of the earth could be disturbed, leading to higher global temperatures. The observation then, of a steady increase in atmospheric CO_2 from an estimated 290 ppm in 1860 to a current value of 335 ppm, has prompted many studies on the possible effects of a continued rise in ambient levels. The estimated increase in CO_2 is the subject of much speculation, but some models predict a possible doubling of its concentration in the next century. The corresponding rise in temperature is also the subject of much speculation, and once again modelling exercises suggest between 1 and 12°C increase in the surface temperature of the earth.

Workshop Activity

Human activities release 20.1 Gt CO_2 yr^{-1} into the atmosphere. There are 600 million cars in the world, of which approximately two thirds are petrol driven. The cars are assumed to have an average mileage of 20,000 km/yr, at 10km per litre (1 litre weighs 0.0007 tonnes). 1.2% of the fuel by weight is emitted in the exhaust as unburnt HC (assumed to be octane). The amount of CO in the exhaust is equivalent to 10.8% of the fuel (and the remainder is CO_2 and H_2O). Calculate the *additional* carbon dioxide loading that is emitted into the atmosphere as a result of fitting catalytic convertors to all petrol driven cars. State the assumptions inherent in your calculations.

The Impact of Catalytic Convertors on the Greenhouse Effect - Tutors notes

Petrol used per year is 8×10^{11} litres, which weighs 5.6×10^{8} tonnes.
1.2% of fuel is unburnt and emitted in exhaust; this weighs 6.7×10^{6} tonnes.
90% of unburnt fuel (assumed to be octane) is converted to CO_2, which yields 1.9×10^{7} tonnes CO_2.
10.8 % of fuel is emitted as CO in exhaust; this weighs 6.05×10^{7} tonnes. 90% of CO_2 is converted to CO_2 which yields 8.6×10^{7} tonnes CO_2.

Using a catalytic convertor on the unburnt HC and CO in the exhaust yields 1.05×10^{8} tonnes CO_2, which *constitutes an increase of ~ 0.5 % in the 20.1 Gt CO_2 emitted by human activities*

Assumptions and other comments:

- Catalytic convertors can use up to 10% more fuel, and thus the amount carbon released would be higher.
- The assumption that petrol is octane is an over-simplification, although impact of using a true petrol content description would be small.
- CO, and to a lesser extent unburnt HC, are converted to CO_2 in the atmosphere, but the temporal and spatial distribution would be different, i.e., the overall impact of catalytic convertors would be even less than the simple calculation suggests.

Capacity Building in Marine Remote Sensing Through Computer-based Modules Serving a Global Network

Dirk Troost
UNESCO
1 rue Miollis
75732 Paris, France

Ian S Robinson
Department of Oceonography
University of Southampton
Southampton SO9 5NH, U.K.

David Blackburn
Bureau International
des Poids et Mesures
Sevres, France

The UNESCO-BILKO project for training in the analysis and application of marine image data has surprised its originators by the worldwide response which it has generated. Using a basic image processing package (BILKO) operating on personal computers, the project provides computer-based lessons to marine educational establishments around the world. The need and requirements for training in marine remote sensing are examined as the context in which this initiative arose, and some of the obstacles to effective training are identified. The evolution of a network of users and producers of lessons is described, the wide range of lesson topics is reviewed, and the international penetration and usage of the programme are analysed. The mechanics of handling the network are also considered. New developments in the programme are described, including a new Windows version of BILKO. It is considered that the most significant contribution of the project lies in the capacity of the growing network to spread knowledge and hands-on training in ocean remote sensing to all parts of the world.

Introduction

In 1989 UNESCO distributed to marine science laboratories and educational establishments around the world, a booklet containing eight self-contained lessons on aspects of satellite oceanography, accompanied by a set of magnetic disks containing a simple image display and analysis software called BILKO, and a number of digital image datasets relating to the lessons (Callison et al., 1989). Responses were requested, including an invitation to suggest ways of utilising the software and of developing its application as an educational tool. As an experiment of computer-based education, distributed unsolicited to a diverse list of recipients, there was no way of anticipating what, if any, response would be obtained. Five years later the project has grown and evolved to a far greater extent than the expectations of its originators.

NATO ASI Series, Vol. I 29
Global Environmental Change in Science:
Education and Training
Edited by David Waddington

Not only was the BILKO software well received as a learning aid for ocean remote sensing, particularly by those whose computing power was limited to basic PC's, but also teachers of marine science responded to the example lessons by producing their own lessons which made use of the BILKO toolkit. As a steady flow of image data and associated lesson material has been offered to UNESCO, so it has been possible, after editing for quality control, to publish these in three further modules. More are in preparation. The resulting body of lessons cover phenomena from all parts of the ocean, using image data from a wide range of sensors, interpreted by local experts from every continent. In this way a network of lesson users and producers has grown almost spontaneously. Whilst the team responsible for the original UNESCO initiative have provided the guidance for this development - the driving energy for it - has come from the community of users itself. International training courses are now making use of the lesson modules. Some of these courses have led to the generation of further lessons, and from one summer school a complete module of eight lessons is in preparation. What started as a small experimental project in computer-based learning is turning into a global network of ocean remote sensing teachers, with a life of its own.

Compared with some other international training initiatives in marine science, the significant impact and influence of this programme has far outweighed the modest resources which initiated it. This paper aims to describe this training programme in more detail and to document its worldwide penetration into marine science educational training establishments. It places the activity in the context of the global training needs for marine remote sensing. However, the underlying objective is to understand why the initiative has been effective, and to identify whether it has wider lessons to teach in general about achieving successful outcomes from international training endeavours.

Training for Satellite Oceanography

(i) Worldwide training requirements in ocean remote sensing

Remote sensing is no longer a specialist technology pursued by a few, but is an observational tool used from time to time by most oceanographers. Consequently all marine scientists need to be trained in the methods of remote sensing. Most need just to learn how to use satellite data, and to be aware of the capabilities and limitations of satellite sensors, although some will need specialist training if they are to contribute to satellite oceanography programmes.

For example, imaging sensors provide a sampling capability which is significantly different from conventional *in situ* ocean measurement methods. They make it possible to acquire "snapshots" of the instantaneous spatial distribution of constantly changing ocean parameters, at high resolution over wide areas. Users of such data need to be taught not only about the measurement methodology and the underlying scientific processes, but they also need to learn how to interpret ocean images. Another important characteristic of satellite-borne sensors is their capability for measuring global distributions of climatically important parameters in a regular, consistent way over many years. Present international marine research programmes to study global ocean productivity and the carbon budget of the ocean, to measure world ocean circulation and the transfer and storage of heat by the ocean, and to examine basin-wide interactions between the ocean and atmosphere, all depend on satellites to provide the spatial context and overview for local and regional experiments. It is evident that effective training must incorporate hands-on experience with image datasets. Pictures in textbooks will not suffice.

The requirement for satellite oceanography training arises in an international context because the use of satellites tends to bring together scientists from many countries into global programmes. Indeed, remote sensing is a particularly useful tool for scientists in countries lacking large resources of conventional oceanographic equipment, because it can provide ready access to large quantities of satellite observations of local seas at relatively low cost. Moreover, it enables isolated ocean scientists to make their own significant contributions to international programmes, for example by obtaining and contributing vital "ground truth" for calibration of satellite datasets. For this reason there is considerable merit in providing training programmes which are international in scope (see e.g., UNESCO, 1988).

Finally, there is also a requirement that education in ocean remote sensing should be offered at a variety of levels. As well as the obvious need to teach ocean scientists at graduate and postgraduate levels, there is also much to be gained from providing the rudiments of satellite data analysis to non-specialists: senior high school students, engineers concerned with environmental issues, planners, politicians, and environmentally aware members of the public. Satellite images have an immediate visual impact and offer an extremely effective way of communicating information about global ocean processes.

(ii) How is training to be provided?

If the need for training in ocean remote sensing is so evident, what obstacles are preventing its provision both locally and internationally, and what can be done to promote it? For a number of reasons, computer-based distance learning appears to be a sensible way to tackle the issue. Firstly there is a need for teachers. Although the importance of satellite data is recognised by all oceanographers, there are relatively few experts with the confidence to offer anything but the most basic of training. Therefore, training materials must be designed for use in a distance learning mode without dependence on local experts. It may be necessary for teachers of marine science to educate themselves before they can teach others, and computer-based distance learning methods enable teachers to learn along with their students.

Since the basis of remote sensing lies in physical science, there is a danger that marine earth scientists and biologists may not be motivated to study it. This would be a great pity since satellite measurements can contribute to all branches of ocean science. Such reluctance can be overcome by offering students a direct hands-on approach to learning, instead of, or in addition to, a standard textbook approach based around the theory of satellite oceanography which does not always inspire scientists to use the technique subsequently.

Unfortunately, the material required to provide hands-on experience of satellite data is not readily available in many educational establishments. Whilst most teachers of oceanography will have access to some satellite images, they are less likely to have easy access to a wide range of digital datasets. But just providing satellite pictures is not enough for students. Pictures tend to lead to qualitative analysis, whereas the goal of the ocean scientist is to make quantitative deductions about the ocean. Training programmes should therefore provide opportunity for the analysis of digital image data as part of the learning process.

The UNESCO Tredmar Computer-based Learning Module

(i) The BILKO image processing toolkit

Given the international requirement for training in remote sensing, constrained by the obstacles identified above, there is evidently a place for distance-learning educational material which can teach students of oceanography and practising marine scientists how to make use of satellite data. With this in mind, UNESCO commissioned the creation of a basic image analysis system which could form the heart of computer-based educational material. The result was BILKO, an image display toolkit software, written by R.D. Callison and M. Dobson of Pharos Scientific, Dundee, UK. It runs on personal computers operating under MS-DOS. To enable it to run on the majority of PCs throughout the world, and hence be widely accessible, it was deliberately kept simple, initially requiring just 640 kB of memory and an EGA display, although a later version makes use of the VGA display.

Another assumption underlying the toolkit was that it would be used by students who had no previous computing experience, so it needed to be very user friendly if its users were to "learn by doing". Therefore, all operations are reached through a simple screen menu, and a contextual help facility is fully built in, aspects which were not as common in PC software six years ago when BILKO version 1 was written. Students should be able to work on their own without supervision, and as far as possible the software was designed to eliminate situations where the machine might fail to respond, and an isolated user would be forced to abandon the work with consequent discouragement.

A fuller description of the software will be found in Callison et al., 1989. Usage of the software has been reported from around the world, and those lacking any access to more powerful image analysis machines have used it regularly as a research tool as well as for its intended educational application. Although its specification now seems dated in relation to recent rapid advances in PC technology, it continues to be widely used. Undoubtedly it was the essential foundation for all the educational activity which has developed from it. However, by itself it was not enough to generate the worldwide response mentioned at the start of this paper.

(ii) Lessons based on Bilko

Lessons must be provided which enable students to use BILKO to learn both the manipulative skills for displaying and enhancing satellite image datasets, and to extract quantitative information leading to scientific interpretations of their oceanographic significance. The demonstration lessons in the first module distributed by UNESCO set a pattern for those subsequently generated by others within the evolving network. The lessons are in printed form, accompanied by the image datasets in digital form on diskettes. An introductory tutorial, several lessons, and disks containing the software and the data make up a complete module. The lessons are structured to be self-contained and accessible by students with little or no local teaching backup. The user is assumed to be isolated and must find all the necessary resources within the module itself. This view of the target recipient underpins the editorial monitoring and review of all lesson material.

Each lesson commences with a statement of its overall aim and specific objectives so that students working in isolation can judge the extent to which they have learned what is intended from the lesson. The lesson text provides background information about the study area and details of the location of the image and how to convert the pixel value on the screen (a number between O and 255) into an ocean measurement. The heart of the lesson consists of a series of steps instructing a student on which functions of BILKO to use and which image files to load. It contains questions to draw out an interpretation and oceanographic analysis of the information contained in the images. Although learning is intended to flow from exploration of the digital image by interacting with the computer, some answers are provided so that students are not left without clear guidance.

(iii) A network of lesson users and producers

The first module was issued with no firm expectations of what might follow. Early exposure of the material at marine science educational workshops soon generated the realisation that lessons based on the software toolkit could just as well be written by recipients as by those organising the project centrally. When the initial mailing was distributed in mid-1989 to UNESCO marine science contacts throughout the world, recipients were encouraged to copy the material and share it with others. They were also requested to feed back comments, and to consider writing their own lessons. There was sufficient positive response of lessons offered for a second module to be planned, and by the time that was distributed in early 1991 (Robinson et al., 1991), it was evident that several more modules would follow, such was the steady flow of new lessons (Robinson et al., 1992).

In this way the network emerged naturally. Because the project operates on a low budget, depending primarily on the goodwill of those directing it under UNESCO's auspices, there was never any vested interest in forcing the programme in a particular direction. Had there been a limited response to the first module, it would have been the last, and the project would have terminated then. It continues to be managed with this philosophy. The user feedback provides the basis for new ideas.

No charges have been made for the modules. It was considered that the benefits to be derived from sales income would not justify the additional administration involved. Also the first module was sent unsolicited, and it is unlikely that the network would have grown so rapidly if modules had had to be purchased. The copyright is owned by UNESCO, and onward sale by recipients is forbidden, but this has not seemed to hinder the sharing and copying of material in a spirit of goodwill. Furthermore, if the project had been set up on a commercial basis, then authors might legitimately look for payment for their contributions which normally represent a considerable investment of time and effort. Instead, authors recognise the benefits to themselves and others of contributing on a sharing basis.

The individual effort in writing a single lesson can be multiplied many times over by receiving the lessons of others, extending the range of training material to cover satellite sensors, oceanic processes, and geographical regions, which would otherwise be beyond the individual's knowledge.

Global coverage achieved by the Project

(i) Distribution of the lesson modules

From the beginning, feedback was considered essential, not only to fuel the development discussed above, but to monitor the extent of use, and thus account for the investment of money, effort, and time by the project team. Without sales figures, the return of a response form by each recipient is important, including those who receive a copy second-or third-hand. To prevent the waste associated with distribution to uninterested recipients, each new module is distributed only to those who have responded from a previous module. Enquiries from names and organisations not on the distribution list are referred, if it is appropriate, to a nearby centre with an established commitment to the network. This has sometimes had the added benefit of encouraging local collaboration and networking between scientists and teachers.

In this way, some regional groups have developed their own momentum. For example, a Spanish-Latin American network has led to the translation into Spanish of modules 1 to 4, and the generation of a new module to be published first in Spanish. In the United States, users are experimenting with the transfer of lessons via electronic mail and data transmission networks. Both of these activities are feeding back into central development of the project.

Based on responses returned to UNESCO, the penetration of Module 3 (the most recent for which the feedback process is complete, is shown in Table 2.

Table 2: Penetration of Module 3

No. of countries receiving Module 3	71
No. of institutions using Module 3	502

It can be assumed that the number of individual users runs in the thousands. It is also pleasing to see that the material is being used in at least 71 countries. Table 3 shows the ten countries which make the greatest use of the lessons. Although the network tends to be most developed in those nations which already have a strong marine science infrastructure, users in the other 61 countries make the network a truly global one, reaching places where satellite oceanography is just beginning to develop.

Table 3: Countries with highest numbers of users

USA, UK, Canada, France, Netherlands, Mexico, Spain, Brazil, Germany, Italy.

(ii) Uses of the lesson material

The reported use of the BILKO image processing toolkit and the associated lessons covers a wide range of situations. Many copies are used individually by scientists, teachers, and students, learning more about satellite oceanography in the distance learning mode envisaged by the authors of the project. Some of the scientists are research scientists in situations with no access to powerful image analysis computers, and use BILKO for simple research applications as well as educationally. The material is also in use by government planning officers and consultants.

Another way in which it is frequently used is by groups of students in classwork situations. The software is either loaded individually onto each of a cluster of personal computers, or else it is placed on a network fileserver. Typically, the groups being taught in this way are undergraduate or postgraduate students of oceanography or coastal science taking a specialist unit in remote sensing. However, some teachers have developed simplified lessons which they offer to first year undergraduates, or even to high school students.

Regional group training activities and summer schools in satellite oceanography and marine remote sensing have also started to make regular use of the material as a simple way of providing hands-on training to support the more conventional lectures. Courses known to have made use of the material include those held in Nairobi, Kenya; La Reunion; Caracas, Venezuela; Enschede, Netherlands; Lepe, Spain; and Rio Grande, Brazil. When teachers running such courses request copies of the material, they are invited to use the opportunity of the course to produce some further lesson material to feed back to the network. This has resulted in a number of valuable lessons produced by groups of teachers and students. In one case, a course at Lepe, Spain, has led to the preparation of a whole module, which will soon be published.

(iii) Lesson Contributors

The variety of lessons published so far is evident from the titles listed in Table 4. They are based on a range of oceanographic phenomena and coastal features

Table 4: Title of lessons

MODULE 1 (July 1989)	MODULE 2 (February 1991)
Images of open ocean sea-surface temperature Thermal images of ocean fronts Coastal Zone Color Scanner data off Iceland Comparison of coincident AVHRR and CZCS data U.K. Shelf seas thermal structure Airborne thematic mapper data - Southampton Water.	Analysis of a SAR image of the Straits of Dover Seasonal plankton distribution - Arabian Sea Coastal discharge - Brazilian Coast Wind-driven coastal upwelling, South Australia Thermal structure of the southern Coral Sea Display and analysis of ADCP (velocity) data
MODULE 3 (February 1992)	**MODULE 4 (September 1993)**
Extent and content differentiation of coastal plumes (Malaysia) Using Landsat for coastal management - mapping and determining major substrate categories Using Landsat for coastal management - vegetation mapping Detection of potential areas for coastal aquaculture using the Indian sensing satellite - IRS 1A Study of Indian lagoonal features using IRS-1A LISS data Temporal and spatial variations in sea-ice concentration in the Southern Ocean Global waves and winds from Geosat Ocean eddy dynamics generated by a numerical model	Imaging *in situ* bio-optical and physical parameters Spatio-temporal evolution of coastal upwelling in the Senegalese and Mauritanian littoral Variation of AVHRR-derived sea surface temperature with satellite zenith angles Seasonal and yearly variation of surface-temperature distributions in the North Sea Analysis of the turbid plume of the Gironde Estuary (France) using SPOT data Detection of sea-surface life with an Airborne Synthetic Aperture Radar Bathymetry prediction from satellite altimeter data

illustrated with examples from around the world, and several different remote sensing methods are covered. There are also some lessons which do not use remotely sensed data, but employ two-dimensional datasets from other sources, since the BILKO software is suitable for lessons based on any image-type data. The authors of the lessons are also from around the world, as illustrated in Table 5.

Table 5: Country of origin of lessons

Lessons published by marine scientists from: Australia, Brazil, Canada, France, Germany, India, Netherlands, Singapore, United Kingdom, USA.
Lessons in press from: Argentina, Chile, Mexico, Peru, Spain, Venezuela

Future Developments

As long as additional lesson material continues to be contributed and the network remains active, it will be necessary to provide editorial coordination of the project. However, personal computer technology is rapidly developing, and hardware is now widely available which is far more powerful than that for which BILKO was originally designed. If the network of lesson users and producers is to be effectively served, it will be necessary to keep abreast of these developments. It has therefore been decided to provide a new version of BILKO which will operate within the environment of Windows 3.1, and Microsoft user interface software, which is now widely available around the world and which is effectively setting standards for personal computer software.

Another exciting possibility is that the lesson material itself can be supplied through the computer, in another window, instead of on the printed page. From this concept flows a number of additional benefits. Hypertext functions can be used to make the lessons more interactive, responding to the needs of the individual user. Linkage between the text window and the BILKO window will lead the student naturally through the lesson, without the need to consciously switch between the two. Because the whole module will be computer-based, the material will be optimised for direct electronic transmission through computer and educational media networks. There will also be scope in future to include multimedia capability into the lessons (e.g., sound and video), and to access data sources other than satellite images (e.g., oceanographic databases on CD-ROM). In this way the network might evolve beyond remote sensing to provide lesson material on oceanography in general. The software development required to make this possible is now in progress.

Discussion and Conclusion

As a worldwide educational initiative, the provision of the BILKO toolkit as the basis of distance learning material in satellite oceanography can be judged a success. It demonstrates the effectiveness, even in the context of advanced technical training, of matching the teaching materials to readily available hardware, rather than attempting to reproduce the sophistication of advanced research tools in the lessons. Beyond the success of BILKO, however, it is the emergence of the user-producer network which has taken the authors by surprise. It may be useful to consider why this aspect of the project seems to have developed a life of its own.

Firstly, by opting for the lowest common denominator in terms of equipment, few potential users have been excluded, and contributions to the developing network have been elicited from all parts of the world without requiring a strong dependence on those nations which have major space and remote sensing programmes. Training in satellite oceanography is to be encouraged in all maritime nations because remote sensing should provide a means for marine scientists from all nations to play their own unique part in the task of understanding earth system science and global change, and in managing the marine environment. It is therefore of significance that lesson material should derive from genuine training contexts in all parts of the world and not be delivered "top down" from outside the network. The BILKO network has grown because it allows local teachers to participate actively as producers, and not just passively as users. It provides opportunities for isolated teachers and scientists to make an internationally valued contribution.

Secondly, the framework provided by UNESCO to administer the network has not sought to dominate its growth, but has tried to provide underpinning support. This has consisted of the original software development and its modification in response to suggestions from the network; the free provision of the modules to committed users, some of whom have acted as local distributors; the provision of a central contact point through the UNESCO TREDMAR office; the establishing of editorial standards for the educational material; and the provision of tutorial material to help lesson creators. The organisational structure has not been established for its own sake, and will remain in being only as long as there is a substantial need to be met in the form of continued demand for educational material and the provision from within the network of new lessons to be edited.

Thirdly, the number of marine science educators in individual countries is usually small, and the concept of sharing educational resources through a network enables them to keep abreast of the rapid pace of technological development in remote sensing. More generally, the network approach offers an effective way to create an international resource base of computer-based learning material in marine science.

However, if the project is to remain effective as the scope of the network grows, it is becoming apparent that it will need to be properly supported. So far it has relied on a small UNESCO budget and the goodwill of the editorial and software development team. Lack of dedicated personnel has caused delays in editing and compiling the modules, and this is frustrating for authors. Provision of additional resources for coordination are therefore necessary to keep pace with the network growth. The core software development also needs to be commissioned at a price appropriate to the standard expected by the network, even though it is hoped that within a windows environment, additional software functionality can be donated by network members. Finally, new lessons require suitable image data, which will not be readily and cheaply available, if space agencies extend pricing structures which are prohibitive for many users of the network. Since the network could be a very effective way to provide skills training for the anlysis of data from new sensors, there is evidently scope for collaboration between the space agencies and the network, including the provision of free data for educational purposes.

References

Callison RD, Robinson IS, Blackburn D, Cracknell AP & Cummings D (1989) Some marine applications of satellite and airborne remote sensing - A computer-based learning module, MARINF/70, UNESCO (Mar. Sci. Div.), Paris, 90pp + disks.

Robinson IS, Callison R, Blackburn D, Boxall S, Garcia C, Monaghan J, Nunes Vaz, Sathyendranath S & Troost S (1991) Applications of marine image data - second computer-based learning module, MARINF/81, UNESCO (TREDMAR Progr), Paris, 85pp + disks.

Robinson IS, Callison R, Carter D, Nichol J, Sheppard C, Sudarshana R, Vifehoff T, Wells N, Blackburn D & Troost D (1992). Applications of marine and coastal image data from satellite, airborne and in-situ sensors - Third computer-based learning module, MARINF/83, UNESCO (TREDMAR Progr), Paris, 101pp + disks 22.

UNESCO 1988, Year 2000 challenges for marine science training and education worldwide, UNESCO reports in marine science, 52, UNESCO, Paris, 129p (available in 6 languages).

Teaching Global Change Topics by Linking Classroom and International Field Research Experiences

Douglas F Williams
Marine Science Program
University of South Carolina
Columbia, SC 29208, USA

Introduction

Under the auspices of the nationally recognized strengths of the Marine Science Program and the Baruch Institute at the University of South Carolina, a special seminar-style course (MSCI 399) titled "Comparative Functioning of Biogeochemical Cycles in Freshwater and Marine Systems" was taught to a team of undergraduates called the Baikal Undergraduate Research Group (BURG). Our hypothesis was that with the proper motivation and mentoring, outstanding undergraduate students could seize exciting opportunities to conduct cutting-edge research, undertake collaboration on an international scale, and develop leadership and organizational skills usually not thought possible at this early stage in their careers.

Goals of MSCI 399

The formal course work of the global change topics course involved readings coupled with weekly written assignments and oral questioning in discussion format to provide undergraduate science majors with foundation knowledge about such topics as global greenhouse processes, global climate change, the biogeochemical cycle of carbon, phytoplankton dynamics and nutrient chemistry, the preservation of sedimentary organic matter, and the physical limnology of large lakes, especially as related to Lake Baikal. In consideration of the students' backgrounds, the program was designed to familiarize students with the challenges of integrating information from a variety of sources and disciplines to attain an higher understanding of complex scientific problems. The extended goal was to link this knowledge with a field research experience in order to gain hands-on exposure to a global change problem. In addition to the formal course work, we also discussed ways for students to achieve critical thinking, establish research priorities, seek funding for research, and experience the process of proposal writing.

NATO ASI Series, Vol. I 29
Global Environmental Change in Science:
Education and Training
Edited by David Waddington

In anticipation that the program would be an on-going, multi-year training effort, the class rank of the BURG students purposefully ranged from first to third year students whereby "senior", more experienced students would be encouraged to assume leadership roles as "peer mentors" as they gained in knowledge, experience, and maturity. In addition to the seminars, we also discussed the nature of collaborative research, methods of seeking research funding, and the importance of disseminating research results through publications and scientific meetings. Special consideration was also given to the exciting cross-cultural aspects of working with Russian limnologists, biologists, and geochemists before, during, and after the expedition. Using a nearby freshwater system (Lake Murray), the BURG students also received pre-expedition hands-on training in the methods of sampling to be employed on the expedition. The classroom and laboratory experiences outlined above were enhanced through a team of faculty mentors both inside and outside USC in the areas of geochemistry, paleolimnology, organic geochemistry, phytoplankton dynamics/sediment trapping, sedimentology/palynology, ecological modeling, and so on. The course outline is given in Table 1.

Table 1. Course Outline for the Marine Science 399 Spring 1994 "Comparative Functioning of Biogeochemical Cycles in Freshwater and Marine Systems"

Topic 1	The Global Climate System
Topic 2	The Carbon Cycle and Global Greenhouse Processes
Topic 3	Introduction to Physical Limnology of Large Lakes
Topic 4	Nutrient Chemistry in Freshwater Systems
Topic 5	Phytoplankton Dynamics and Nutrient Cycles
Topic 6	Limnology and Phytoplankton Productivity of Lake Baikal
Topic 7	The Biodiversity of Lake Baikal
Topic 8	The Origin of Lake Baikal and Rift Basin Sedimentation
Topic 9	Principles of Ecological Modeling
Topic 10	Introduction to Organic Geochemistry of Lacustrine Sediment
Topic 11	Evidence for Paleo-pCO_2 from Marine, Lake and Ice Core Records

The Field Research Experience

To link classroom discussions of Global Change with the field research aspect of the program, we scheduled 3 weeks of ship time on Lake Baikal in May-June 1994 to coincide as closely as possible to when the frozen surface of Lake Baikal is melting (Sitnikova et al., 1984; Bolgrien, 1993). As the surface waters are exposed to atmospheric CO_2 after 4 months of isolation and as the microscopic phytoplankton flourish, we wanted to determine how the carbon reservoir of Baikal was affected and what geochemical signals in both the organic and inorganic carbon reservoirs of the surface waters are recorded and preserved in the underlying sediments. Previous work with our Russian colleagues has shown that silica and organic carbon from diatoms are dominant components of modern Lake Baikal sediments. Previous work on the carbon cycle in a small Swiss lake by Hollander and McKenzie (1991) has shown that the organic carbon from phytoplankton faithfully records the seasonal changes in the concentration of CO_2. With these results, the authors proposed that the organic record of lake sediments must act as a monitor of past changes in atmospheric CO_2 (paleo-pCO_2). By verifying the response of Baikal's surface waters to changes in atmospheric CO_2, we wanted to define some of the parameters for determining whether a paleo-pCO_2 record is represented in the sediments of Lake Baikal.

The BURG students with their American and Russian faculty mentors conducted the sampling at 14 stations (Fig. 1; Table 2). The exact number of stations and the depth of sampling were determined by the depth of maximum

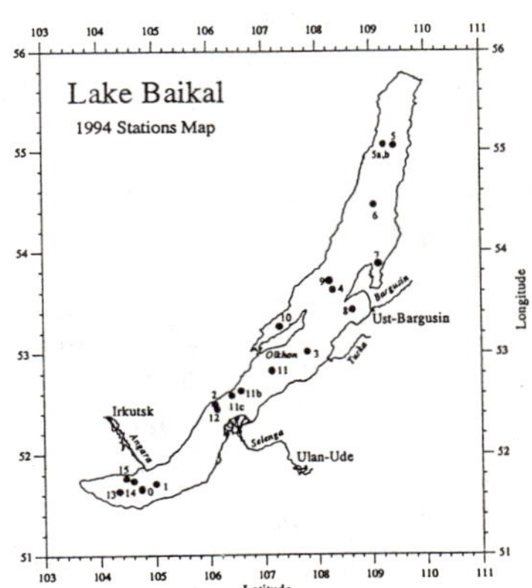

phytoplankton. In addition we attempted to document the nature of the organic materials falling through the water column (sediment traps) and arriving at the lake bottom (box coring). Aboard ship, one or two students were responsible for a particular sampling task and subsequent analytical work. All students were familiar with our reasons for conducting each sampling task and understood that all data would be pooled, discussed, and interpreted as a group in consultation with associate faculty of the project. It was anticipated that all of the students would continue working on the samples as part-time assistants for the summer and fall 1994-spring 1995 semesters. Prior to the expedition, a detailed analytical plan that took into consideration existing facilities and expertise was decided jointly by the students and our Russian colleagues.

At least two BURG students are using data from the expedition for their senior theses in the South Carolina Honors College. Two students remained in Irkutsk for the remainder of June-August (ten weeks) to participate in additional expeditions and to work with their Russian mentors at the Limnological Institute and Institute of Geochemistry. We are now working to utilize existing satellite imagery and remote sensing to track certain variables of Lake Baikal related to the *in situ* stations. Our eventual goal is to obtain an annual time series of the changes in phytoplankton flora (biomass, diversity, etc.), nutrient chemistry, and geochemistry of the organic and inorganic carbon reservoirs

Table 2. Sampling Protocol for May-June 1994 Baikal Expedition

Type of Sampling	Analytical Determinations
1. plankton towing using nets arrayed to sample the maximum phytoplankton abundance of the upper 250 m	special identification biomass measurements stable isotope analyses
2. water column sampling using Niskin bottles	total dissolved CO_2 alkalinity, pH nutrient chemistry (nitrate/phosphate)
3. water column sampling of particulate organic carbon (POC) by filtering pumped water	stable isotope analyses
4. sediment trap deployment	total organic carbon/nitrogen content stable isotope analyses biogenic silica content inorganic particulate content
5. box coring of uppermost surface sediment	total organic carbon/nitrogen content biomarkers (organic geochemistry) stable isotope analyses biogenic silica content
6. CTD deployment	pore water geochemistry profiles of temperature, pressure, light transmission, dissolved oxygen

Acknowledgments

I personally want to thank the many Russian and American individuals and faculty mentors who helped make possible the formation of the Baikal Undergraduate Research Group: Sarah Fowell, Oleg Timoshkin, Nina Guselnikova, Nina Bordeenko, Evgeny Karabanov, Valentina Sidelova, Mikhail Grachev, Mitchell Colgan, Dwayne Porter, Steve Davis, Robert Thunell, Eric Tappa, Tom Chandler, Bill Johnson, Willard Moore, Tim Shaw, David Hollander, Jim Morris, Hank McKeller, Jim Collister and Dave Wavrek. Of course, none of the expedition's success would have been possible without the effort of Captain Oleg Kalinin and the crew of R/V *Vereschagin*. On behalf of BURG, I also gratefully acknowledge financial assistance from the Marine Science Research Fund of the Provost's Office, Dr. William Murray and Samuel Freeman Charitable Trust, the Explorer's Club Research Fund, the REU program of the National Science Foundation, the South Carolina Honors College, the Belle W. Baruch Institute for Marine Biology and Coastal Research, and the USC Marine Science Program. In addition I thank the Limnological Institute (M. Grachev), the Institute of Geochemistry (M. Kuzmin, T. Bounaeva, and E. Almukhamedov), the Baikal International Center for Ecological Research, the United Institute of Geochemistry, Geophysics and Mineralogy of Novosibirsk (V. Efremov and N. Dobretsov), and all of the Siberian Branch of the Russian Academy of Science for many forms of assistance.[1]

References

Bolgrien DW 1993 Delineation of the Hydrodynamics of Lake Michigan and Lake Baikal Using Satellite-Derived Surface Temperatures. PhD Dissertation, Biological Sciences, Univ. Wisconsin, Milwaukee

Hollander DJ and McKenzie JA 1991 CO_2 control on carbon-isotope fractionation during aqueous photosynthesis: a paleo-pCO_2 barometer. Geology v 19, p. 929-932

Sitnikova GV, Furman MS Yanter NN 1984. Remote and ground-based methods for researching the dynamics of ice processes in Lake Baikal. In: Remote and Ground-based Studies of the Dynamics of Natural Processes in Siberia. Inst. of Geography, USSR Academy of Sciences, Irkutsk, p. 72-81

---

1 USC is a rounding member of the Baikal International Center for Ecological Research, along with the Siberian Branch of the Russian Academy of Sciences, the Royal Society of London, the Belgium Academy of Natural Science, the Japan Baikal Research Association, and the Swiss Federal Technological Institute.

Global Change: A Context for Teaching Science

Joseph P Stoltman
Department of Geography
Western Michigan University
Kalamazoo, MI 49009, USA

David J Waddington
Department of Chemistry
University of York
York, YO1 5DD, U.K.

Introduction

This workshop is concentrating principally on the training of chemists who will be concerned with global change, either in research or management. Such scientists usually come from single-subject honour degrees in physics, chemistry, earth science, and biological sciences, yet they are expected to understand and appreciate the role all sciences contribute to the understanding of the key issues in the environment, which are known under the umbrella title 'global change'.

The problem is exacerbated by school and junior university courses in which the sciences are also taught in terms of specialisms using boundaries defined a century ago. Further, the content is often delivered as a series of received truths, in terms of simple models. Environmental concerns have therefore very little impact on many school and university science courses. This leads, in turn, to many students seeing science as irrelevant to their needs and being attracted to other subjects in the curricula which *appear* more interesting and suitable for their careers.

Education in Global Change Project (EGC)

The Education in Global Change (Stoltman and Waddington, 1994) is a new project sponsored by the International Council of Scientific Unions[1] for students in senior secondary schools and colleges (ages 16-20). EGC was created by a group of science educators drawn from France, India, the United Kingdom, and the United States, with considerable help from scientists involved in the International Geosphere-Biosphere Programme : A Study in Global Change, to ensure that the materials were accurate and fair.

---

1 *International Council of Scientific Unions* has a membership of scientific academic and research councils and international scientific unions (for example, the International Union of Pure and Applied Chemistry, IUPAC).

NATO ASI Series, Vol. I 29
Global Environmental Change in Science:
Education and Training
Edited by David Waddington

These materials are put together in six units:

The Changing Atmosphere
Clues from the Past: Glimpses of our Future
The Global Cycle
Population and Land Use
Oceans
Remote Sensing: Windows on Global Change

Each unit has a series of **student activities**, which can be used to introduce important scientific principles, principles such as met in courses at this level in thermodynamics (energetics and equilibria), in reaction kinetics (catalysis and mechanisms), free radical chemistry, chemical analysis, the electromagnetic spectrum.

The principles are set in the context of key contemporary environmental issues, which, in turn, help students towards the understanding of issues such as the protection of the ozone layer, the enhancement of the greenhouse effect and the destruction of the rain forests.

We want, at the same time, to show that science is a force for good and that scientists have an important role in enhancing the quality of our lives. To show this, students must be able to appreciate what advice scientists can give to decision makers, be able to judge the quality of that advice and appreciate what effect the advice, if accepted will have on everyone. It is important that students are able to make up their own minds about these and issues yet unfocussed. Thus, many of the activities have been made as open as possible to allow students make up their own minds about the conclusions it is possible to make from the evidence.

To summarise, the activities are designed to:

- cover key scientific concepts;
- illustrate science in the context of important environmental issues;
- encourage student participation;
- encourage skills such as data interpretation, decision making and communication; and
- illustrate the positive role played by scientists in society.

The units can be used in different ways:

- as a whole unit, occupying about 10-12 hours
- in existing courses, as single activities to illustrate a specific scientific principle or environmental issue by reassembling
- as items from more than one unit to produce a unit, specific to school or college.

Each unit begins with an **introduction** for the teacher which includes a synopsis describing the student activities. These are brief descriptions so that teachers can readily select those activities they wish to use for their courses. We do not expect the activities to be used exactly as printed. Many teachers will wish to tailor them for their own purposes. The **activities** include laboratory experiments, data collection, data interpretation, literature surveys, short talks, and writing. There is a minimum of teacher-led work, with most time devoted to the teacher assisting the students in discussing procedures and judging the significance of their findings. EGC is *not* extra work. Rather the activities provide *alternative* means for the teacher to introduce or illustrate principles which are in their courses in a new and exciting global change context. Since some teachers are themselves likely to come across the global change context for the first time, great care is taken in the units to ensure they are given adequate background information in the **Teachers' notes.** Here are also found the answers to questions and discussion suggestions, hints about setting up experiments and suggestions for teaching strategies.

Examples of Activities

(i) Key scientific concepts

In one activity in *The Changing Atmosphere*, the greenhouse effect is used as a way of introducing the properties of the electromagnetic spectrum. The students design and carry out an experiment to illustrate blackbody radiation in which they use powdered graphite, some oil and two thermometers. In the same unit, they collect and then interpret absorption data for a series of gases, including oxygen, nitrogen, methane, carbon dioxide and CFCs. The activity introduces them to the properties of molecules which lead to different absorption patterns and on to the idea of the atmosphere functioning as a greenhouse window. The properties of the electromagnetic spectrum are reinforced when its role is examined in another activity, in *Remote Sensing: Window on Global Change*.

In *Global Carbon Cycle,* students carry out a series of basic experiments using a bottle of soda water. The closed bottle is treated as a system at equilibrium and when the cap is removed, the dynamic equilibrium is disturbed. A series of questions and experiments unfold. How much carbon dioxide will water hold? What happens when the system gets warmer? Thus the solubility of gases is introduced related to an atmospheric context. The reaction chemistry of carbon dioxide is also introduced in this way.

(ii) Student participation

We have designed the activities so that students participate in their design and interpretation, illustrating many of the dilemmas facing scientists. *Land Use and Population* examines the human dimension of global change. It begins with an analysis of population change and the demands populations place on the global environment both for resources and disposal. The contribution of land use to the production of several global change elements, such as carbon dioxide and methane, are analysed as case studies. Care is taken to examine cause and effect relationships, and to develop and test hypotheses that require the use of statistics and information from sciences other than geography.

In *Clues from the Past: Glimpses of the Future,* students use actual data to determine long term changes in the atmosphere, and evaluate their importance in projecting what might happen in the future. The data include evidence such as the size of tree-rings, pollen and, perhaps most dramatically, samples of ice drilled deep in Antarctica. The activities include a laboratory experiment and game designed to give practice in interpreting data.

Remote Sensing: Window on Global Change relies upon satellite and remote imaging technology to analyze the changes that have occurred with land cover in several regions of the world. The students are given the opportunity to examine the images, make measurement and use the image as a source of data, process the data and test hypotheses related to both the temporal and spatial implications of global change. The use of temporal data permits the students to make predictions, and to check those predictions against the uncertainty that accompanies scientific investigation. A case study is presented of Rondonia State, Brazil. This real life laboratory is revealed *via* remotely sensed images that are used to measure the increasingly greater amounts of deforested land between during three time periods. The students observe and analyze the changes, relate the changes to other regions of the world, and begin to measure the cumulative effects of human action on land cover and the relationships between people and environment.

In *Global Carbon Cycle,* students interpret data, and construct, both qualitatively and quantitatively, a carbon cycle that includes considering the limits of data and the types of errors scientists must consider.

In *Changing Atmosphere,* students learn about the reactions of nitric oxide and ozone. Then they are asked to predict what effect a new fleet of supersonic aircraft may have on the stratosphere since both nitric oxide and ozone will be present. They have to decide what data are required and what assumptions they need to make. They are faced with a real problem, such as scientists face when they predict the effects of scientific and technological changes on the environment. For example, this activity, based on the work of Harold Johnston, Douglas Kinnison and Donald Wuebbles (1989) gives students a salutary lesson in the difficulties of looking far into the future to explore and predict compounding effects. Finally, the students are asked to write a short account for a newspaper read by a general clientele. There is nothing to compare with the difficulty of trying to explain science to a non-scientific audience, and the student soon realizes that to do so, you have to have a real mastery of the science yourself!

The *Education in Global Change Project* has brought scientists and science teachers together to collaborate on teaching resources for schools and colleges which illustrate key scientific principles and, at the same time, show the essential role of scientists in society. This, combined with the international structure of the project, demonstrates that science and the global environment have no national boundaries.

References

Johnston HS, Kinnison DE and Wuebbles DJ (1989) Nitrogen oxides from high-altitude aircraft: An update of potential effects on ozone. J Geographical Research 94: 16351 - 16362

Stoltman JP and Waddington DJ (eds) (1994) Science for understanding tomorrow's world: Global Change. International Council of Scientific Unions.

Table 1
Education in Global Change Units

The Global Carbon Cycle
At the heart of the problem of global change lies the role of carbon dioxide and its fate in the atmosphere. This unit includes authoritative data on the various reservoirs and fluxes in the global carbon cycle. Activities require students to interpret the data, to carry out simple experiments and to discuss some of the global change consequences of the cycle.

Remote Sensing: Window Global Change
Remote sensing plays an important role in detecting changes in land cover, types of vegetation, and ocean surface conditions. This unit includes exercises in the use of remote sensing techniques ranging from aerial photographs to satellite imagery to monitor changes occurring on the earth's surface.

The Changing Atmosphere
The unit gives students a firm understanding of the effect of trace gases produced naturally and by human activity, on the stratosphere. It includes the greenhouse effect and the role of the ozone, methane and chlorinated hydrocarbons in the troposphere and stratosphere using a series of data interpretation exercises.

Clues from the past: Glimpses of our future
The unit introduces students to a range of natural archives which can give scientists clues about climate change in the past. Data analysis and practical exercises show students how scientists interpret the results of studying tree rings, ice cores and pollen grains. The units end with activities to underline the significance of these studies both for scientists interested in climate modelling and for policy makers in society.

Population and land use
Population and the activities of people are two significant variables in the investigation of global change. Global population has increased dramatically during the 20th century and growth projections suggest continued increases into the next century. Changes in population have an important effect on land use in many parts of the world, with more land being used for agriculture in particular. The effects of increasing population and the contributions of changing land use to global change are investigated.

Oceans

The importance of the oceans when considering global change are illustrated in two sections. One, entitled "Blooming Algae", gives an overview of the way in which algae in the oceans interact with carbon dioxide in the atmosphere, during which ideas on positive and negative feedbacks are introduced. In the second, "Salty Sinks and Sources for Carbon Dioxide", students examine data giving possible carbon dioxide production and absorption in the oceans.

SECTION 4: WORKSHOP DISCUSSIONS

Education and Training in Global Change Science
A review of the working groups

John Boardman
Environmental Change Unit
University of Oxford
Mansfield Road
Oxford, OX1 3TB, UK

Lynne Carter
Graduate School of Oceanography
University of Rhode Island
Narragansett, RI, USA 02882

Rudolf de Groot
Centre for Environmental and Climate Studies
Wageningen Agricultural University
PO Box 9101
6700 HB Wageningen, The Netherlands

David Waddington
Department of Chemistry
University of York
York, YO1 5DD, UK

There were three working groups, one concerned principally with early undergraduate courses, another with senior undergraduate courses, and the third with postgraduate courses. Each group was asked to structure its discussion along the following lines:

What are the significant objectives of courses in global change science at the level of education being discussed by the group?
Where are we now?
Where should we be going?
How do we get there?

The groups worked independently, coming together for plenary sessions to report on progress and to give advice to one another on possible discussion points. Each will be described and, at the end, there is a synthesis.

WORKING GROUP 1: SENIOR SECONDARY AND FRESHMAN UNIVERSITY COURSES

Working Group 1 addressed the role of global change science in senior secondary and early college courses. The focus was on introductory level courses concerned more with breadth than with depth. The courses are intended for all students, not only those planning a career in science, but also those planning non-scientific professional careers.

NATO ASI Series, Vol. I 29
Global Environmental Change in Science:
Education and Training
Edited by David Waddington

The purposes of these introductory courses are to familiarize all students with the earth system concept and the role of humans in that system and to contribute to the development of responsible citizenship by

- improving their scientific understanding (content) related to global change issues,
- teaching them how to use scientific evidence in addressing issues in global change,
- enabling the application of knowledge about the earth system to the evaluation of public environmental information, and
- encouraging them to appreciate personal behavior and increased participation in public policy related to global change issues are necessary.

Where are we now?

A growing proportion of the population is interested in understanding a wide range of global environmental issues. Interests range from a future as an environmental scientist to that of a concerned citizen wanting to participate in the increasingly technical public policy debates. The educational needs students in senior high school and the first year in college or university are being met, at least in the US, through offering an introductory or survey course of one or two semesters that addresses the wide range of interactions between humans and the natural environment. There are fewer examples from other countries. However, even in the US, there are a number of problems that face the introduction and maintenance of these introductory courses. Issues and questions include: interdepartmental 'turf' and support; granting science credit to students when the course also discusses human dimensions; who should be required or allowed to take such courses; developing an introductory course with available expertise; departmental responsibility for cross-disciplinary courses; and course focus for a mixed audience (e.g., science and non-science majors).

Many of these potential difficulties must be handled course by course, and university by university. The mixed audience issues, however, were generic enough that some comments were offered by the participants which is described in the next section.

Where should we be going?

The working group felt that all educational institutions should play a role in increasing staff and student awareness of the interactions between humans with the environment. The information to achieve this should be available at all levels of education. While the focus of this report is on formal education, the group considered that informal education related to global environmental issues should be offered to all citizens at all levels through a variety of outreach efforts that hold particular implications for teachers in primary and secondary education.

The following *Overview of an Introductory Course in Global Change* represents the sort of course that could become a regular offering from many educational institutions. The working group recognized that it is impossible to address all of the issues that could fall under a heading of global change in any one course, but that any survey course aimed at the senior high school or early undergraduate student should include four important perspectives. The four perspectives are earth system processes, interactions, the role of humans in the global system, and tools and approaches. The themes and order of perspectives chosen will depend on the expertise available at the educational institution as well as the structure of the course and its function.

Overview of an Introductory Course in Global Change

Specific aspects of the four perspectives important to any introductory course on global change are:

1. ***Earth system processes:*** *links between and processes in the atmosphere, hydrosphere/cryosphere, biosphere, and lithosphere*

   Depending on the global change/human interaction issues emphasized in the course, the basics of different aspects of the earth system involved must be understood. For example, if ozone depletion is the issue being addressed, then atmospheric circulation must be understood; or if global warming is the issue, then oceanic circulation, biospheric interactions and the greenhouse effect must be prominent.

2. ***Interactions:*** *natural and anthropogenic influences on element cycling in the earth system*

   The important focuses of this course are the interactions between the human system and the physical sea/land/air system. The emphasis in this category should be towards an interdisciplinary (not a multidisciplinary) approach to learning. Using global warming as an example, natural and anthropogenic contributions to greenhouse gas build-up (i.e., volcanic activity, plants, such as rice growing which gives off methane, and fossil fuel use) and the feedbacks in the system need to be understood and discussed.

3. ***Influence of humans in the global system:*** *anthropogenic influences, impacts, and responses*

   Specific issues that contribute to and are impacted by global change can include population, energy, resources (depletion, distribution, recycling), acid rain, desertification, global warming, ozone loss, land cover changes, biodiversity, sea level rise, water distribution, socio-economic issues, equity concerns, and response options including individual and collective, economic, national and international, and technical.

   The issues that are chosen as a theme or a focus for any specific introductory course could be any combination of the above. Again using the example of global warming, one might include a discussion of the economics of fossil fuel use versus alternative energies, and all of the social and environmental implications of our energy choices.

   The definition of what issues are considered global environmental change are variable, thus the list of choices. The issues chosen will determine which earth processes are included, which interactions are focused on, and therefore the teaching strategies, such as laboratory or field work and role playing, to be utilized in the course.

4. ***Tools and approaches***

   In teaching this interdisciplinary course to the chosen audience, it is important to utilize a number of approaches, one of which addresses the understanding of tools that scientists use to collect data. Familiarity with appropriate scientific tools is an important perspective in understanding uncertainty of scientific data and the resulting information. It was deemed crucial that participants have an historic perspective, as well as an introduction to such scientific techniques and viewpoints as

   - scientific method and experimental techniques
   - remote sensing of the environment
   - Geographic Information Systems (GIS)

- computer models and simulation of environmental change
- chaos theory and limits to predictability
- earth history, palaeoclimatology, and past environmental change
- analysis of uncertainty and perception of risk

The choice of what technologies to discuss, again relates to the topic chosen. In global warming, how carbon dioxide is measured in the atmosphere, past and present, would be important to understand.

The wide range of potential audience for this sort of an introductory course implies a disparity of backgrounds, which in turn suggests a difficulty in choosing appropriate teaching focuses and strategies. Some of the options suggested for dealing with different groups in the same audience include:

- additional work or sections for advanced students in their own speciality and interest, and
- use of creative teaching methods, such as role playing exercises, interactive computer technology, team or individual projects, and other methods which allow active participation by students in their education. Some examples are given in the papers.

How do we get there?

The approach to getting a course / unit like this offered in an institution will be specific to each institution. However, there are a number of models that have been employed with various levels of success.

Top down: A senior faculty member initiates or is willing to support strongly junior colleagues to mount a new course.

Bottom-up: Strengthen or extend existing courses. This can lead to the evolution into a whole new course. One discipline may first show by example; then others copy it by extending their courses. Eventually there should be a groundswell of support for a fresh new coordinated course in global change science.

Staple gun[1]: Use existing courses or pieces of courses. This is often used for a unit or grouping of courses and is easier if the university has a modular system, such as that utilized in the North American model. If a university has a tight cellular structure as in many European universities, there are considerable staffing problems for new courses.

---

1 Staple gun - a vivid metaphor much used in our discussion, implying the collation of a new course by stapling together modules from pre-existing courses. (Schneider 1994)

The working group also wanted to comment on the broader aspects of the goal for all educational institutions to be actively pursuing programmes for all levels, aimed at increasing awareness of our relationship with the environment. A variety of strategies could be employed to achieve that goal. One strategy could be to develop a series of focused offerings, for example, a specific course for school teachers, businesses or the public. Another strategy could be employing distance-learning methodologies (i.e., electronic, TV, video) for those unable to attend. Finally, an important group to reach are those scientists that have the information but have no experience in offering other than technical programmes.

A university group could arrange special sessions at professional meetings to provide available resources to encourage outreach efforts by its faculty and staff.

WORKING GROUP 2: SENIOR UNDERGRADUATE COURSES (3RD AND 4TH YEARS) IN GLOBAL CHANGE SCIENCE

Working Group 2 addressed the role of global change science in advanced undergraduate courses (for 3rd and 4th years). The courses considered were those aimed specifically at science students with a good, though discipline-oriented, knowledge of science at the freshman level. The objectives were not dissimilar to those from Working Group 1: namely that students in global change science should

- improve their scientific understanding in contexts related to global change to enable them to be prepared for studies at postgraduate level and for professional careers
- become aware of key issues in global change
- be trained to use systems analysis and models, and risk analysis
- be given as wide a view as possible of the impacts of global change on society through evaluation of implications of global change scenarios and therefore development of preventive policies to reduce risks.

Where are we now?

The development of programmes that cater for students in advanced undergraduate courses are at very different stages, not only in different countries but also between institutions. It is perhaps the stance taken by the institution towards innovation in this field, which has the greatest impact on development of such programmes.

Different approaches are being tried to develop interdisciplinary courses, and most of the current work is in producing courses within single disciplines, which have a heavy bias towards global change examples. The Lisbon (Camoes and Neves 1994) and York, UK (Anastasi 1994), models are of this character. The U.S. system, in which faculties and schools hold relatively more influence than individual departments, allows for a multidisciplinary approach, drawing on existing courses in different faculties and departments to provide the content. Examples in this category include courses at the University of Colorado (White 1994) and at Stanford University (Schneider 1994). The drawback to this system is that there is little or no control on content and inevitably this leads to a lack of coherence. Nevertheless, it is a system that introduces students to a subject of global change science and enables the faculty to build up experience and expertise in this sector of science education.

Where should we be going?

There is considerable evidence that interdisciplinarity in science education is seen by many to cut across what are regarded as deeply held, established tenets. One such tenet is that it is important at this stage of education to give students a good grounding in discrete scientific disciplines and that there is only time to do this in one, or at the most two, disciplines. Secondly, interdisciplinary courses cut across established institutional funding arrangements to the perceived detriment of existing departments. Other difficulties are related to associated research where full credit is not given to achievements that transcend boundaries.

The group did not feel that any difficulties provided insurmountable problems. While agreeing that it is important for future global change scientists to have a thorough grasp of important concepts of the different sciences, it is nevertheless possible, given goodwill, to ensure that concepts, which are irrelevant to global change science training are eliminated. Secondly, by using examples and global change science while teaching fundamental scientific principles, time is saved and courses become more lively, relevant and accessible.

The 'staple-gun approach' has many limitations and the creation of specific courses is important if measurable progress is to be made. This requires a dedicated organisational structure including tenured faculty within an existing department or institution, but whatever structure is chosen, it must be flexible in its approach. The group did not develop a general outline for a 'global change course', believing that at this level it is dependent, at least in part, on the experiences and expertise of the faculty. However, judging by the courses described in presentations, there is a common-core which can be abstracted.

How do we get there?

Again, this question was mainly discussed from the 'institutional' point of view in order to provide a stable base for the development of truly interdisciplinary global change courses.

To achieve institutional stability for interdisciplinary global change courses, a so-called 'successional model' was suggested (Schneider 1994) with the various steps, starting with multi-disciplinary seminars and the instalment of a 'programme committee' to draft staple-gun courses leading to the development of new 'core' courses. Simultaneously, new discipline tracks and interdisciplinary senior seminars should be developed, and the possibilities for co-term MSc degrees widened and encouraged. Ideally, all this could lead to the development of new departments/schools or centers/institutes. It was noticed that this model is not truly successional and that there may be 'jumps' along the road.

The group split their discussion into two parts

a) the identification of the main obstacles, followed by
b) an attempt to describe some possible solutions.

(a) Some problems

Some problems concerned with structural administration of the institution may be

- a lack of 'political muscle' within a single department or in a group to overcome institutional inertia
- financial constraints
- a lack of credit given to the individual for teaching in interdisciplinary courses organised by another department.

(b) Solutions

- Depending on circumstances, the original ideas may come either from junior faculty or from faculty leaders. Regardless of which, in order to get courses started, course providers should discuss complex issues. This requires an experienced and well-respected leader and a group of analytic thinkers. Although not ideal, staple-gun courses are a realistic, and probably in practice, an essential first step.

- Marketing and networking: because we are advocating a new approach, the idea has to be sold to existing institutions, both to develop the course content which will mean trespassing on existing disciplinary and institutional borders, and to obtain structural funding. Demonstration projects and field-excursions were suggested as possible activities for demonstrating marketing and networking strategies.
- The link between research and teaching is as essential in global change science as in other science courses, if not more so. Credit for interdisciplinarity in research has to be put on the same footing as that for conventional research.
- Again, as in other courses, it is important to be sensitive to the future needs of students, whether for direct employment after the course or in pursuing their education through higher degrees.
- There is need for regular evaluation and feedback on the course, the teaching, and on student progress.
- Summer courses/schools and exchange programmes are important at this educational level to provide off-campus ('real world') experience in global change science. They could be regionally oriented, for example, short courses at the Arctic Center, University of Lapland (Moore and Lange 1994), or more thematic or problem-driven, such as excursions or short courses to sites where the relation between climate change and desertification is studied, in Spain (Medalus) or to the permanent research site in a boreal forest in Norway (Climex-experiment). The work at Wanginengen University also deserves special attention (de Groot 1994), as does the course on Climate and Environment, which is planned by the Open University of the Netherlands in collaboration with the Open University in the UK and the Wanginengen University.

WORKING GROUP 3: POSTGRADUATE PROGRAMMES IN GLOBAL CHANGE SCIENCE

Working Group 3 addressed the issues involved in designing postgraduate programmes, particularly courses leading to masters degrees, whose major component is course work and which allow for specialisation in global change science.

The purpose of such courses is to provide the necessary subject content and practice to develop skills which will enable students to contribute to the teams of experts who are attempting to understand problems in global change science.

Heavy emphasis was laid on the point about teams as it is inconceivable that global issues will be dealt with in organizations by solitary individuals. Although global change science may concentrate on human-induced perturbations of the global system, inclusion of background trends (natural variations) is necessary.

The programme will accept students from different academic backgrounds. They may be from single subjects (e.g. biology, chemistry, physics), from interdisciplinary courses, or from non-science programmes (e.g. economics, anthropology).
The end product wanted and the motivation of students will vary. Some participants will wish to pursue research; others will be professionals, training or re-training in global environmental science; others will be motivated by curiosity. This, in turn, should lead to a variety of different programmes and, within the courses, to a flexible approach to programme design.

Where are we now?

As in the case of the other working groups, interesting differences between national systems were noted. Particularly relevant in the context of postgraduate programmes is the greater reliance in US systems on course textbooks, in comparison with systems in Europe where they play a lesser role. Taught elements play a larger role in US courses, even in doctoral programmes. However, there is a small number of masters courses which cater for global change scientists.

A programme's very novelty will lead to a large burden on the staff promoting it. There is little experience to draw on. Each activity is likely to have to be thought through afresh, teaching and learning methods questioned, new resources produced, and new assessment procedures designed. The normal problems facing faculty in setting up courses are thus magnified. Although it is very likely that programmes will be put together, in the first instance, by the staple-gun approach, there will be constraints, in some cases severe, on what can be achieved by staff who are often discipline-based. Because the courses that are being designed are new, the constraints, observed by Working Party 2, are likely to be exacerbated.

Two examples are to be found in the UK. The Oxford MSc degree course (Boardman 1994) is being created from existing sets of courses, and is based in a research unit. At York (UK) the MSc degree course in Environmental Economics and Environmental Management, while the brainchild of the Department of Biology and Economics, has as its core a new Department of Environmental Economics and Environmental Management with students also taking relevant courses in Biology and Economics. As has been stated for other courses, this one is designed with two goals in mind: (i) to provide a professional qualification for those intending to work or already working in the area of environmental management, and (ii) to prepare those who wish to enrol in a research degree in environmental economics and environmental management. The course combines insights from two key disciplines: economics and biology. Drawing largely from the sub-fields of environmental economics and ecology, it offers a set of skills not at present available in other courses. From the field of ecology, it offers insights into the functioning of ecological systems, and the nature of the ecological feedback effects of economic utilization of those systems. From the field of environmental economics, it offers insights into the optimal utilization of scarce environmental resources, with and without markets. The combination of the two enables students to learn how to incorporate environmental feedbacks into the economic decision-making process in a way that satisfies the goals of both resource managers and resource economists. To be eligible for direct entry into the MSc programme, applicants must normally have studied economics and biology to an advanced level in their first degree. Students who do not satisfy these requirements will normally be required to do a foundation year - comprising a set of upper level undergraduate courses tailored to the individual needs of the student (University of York, 1995).

Where should we be going?

A masters course which is principally concerned with course work is probably the most appropriate model. It will be an advantage if the course is constructed so that it can be taken either full- or part-time over a larger time span. The latter mode will be an advantage to those students in employment.

A main issue in course design is the provision of elements of both breadth and depth. A suggested balance which gives equal weighting in time and assessment is outlined below.

Section 1 *Breadth* : students take all sections of this part of the course. Details are eschewed, as it will depend on the precise aims of the course and the resources at the institution. However, in general terms it should cover aspects of the following broad areas: earth as a system; causes of change; effects of change; socio-economic consequences; response; human behaviour.

Section 2 *Depth* : students choose options in discipline areas of special interest to them such as environmental law, environmental economics, and advanced scientific-based courses.

Section 3 A research *project* leading to a dissertation

How do we get there?

There are problems and constraints to consider as the course is being designed and during its early years. The problems are similar to those described by Working Group 2. The development of the course may be constrained by

- the expertise of staff already in post
- the background knowledge of students
- lack of funding
- the lack of developed and well-tested teaching resources, e.g., software and datasets books
- the necessity to develop appropriate student assessment techniques

There are the institutional problems which include

- academic inertia - a particular problem for interdisciplinary programmes and the staff involved in these programmes
- structural inertia (institutional barriers between departments contributing to interdisciplinary programmes)

Once again the solutions are largely similar to those discussed by other groups. The following were considered essential.

- Enlist international support for an exciting new programme, and especially obtain financial support for students from less-developed areas

For this it is helpful to be able to show that the programme already attracts a large number of students

- Develop new teaching resources, such as software, books and multimedia resources, and, where possible, collaborate with other institutions by sharing materials developed in-house
- Limit the number of options at the start and expand the number as the course becomes established
- Strive for flexibility in the design of programmes, so that Section 1 of the core can also be used as an *outreach* course.

A synthesis

It is unsurprising that the discussions overlapped and several salient points are common to all levels of education:

1. The introduction of global change science courses competes with existing deeply-entrenched courses in departments with long-established disciplines. Such departments have relatively stable infrastructures of funding and staff.

2. There is an unresolved (and perhaps unresolvable) debate, within the global change science area, on the relationship between the depth and the breadth of understanding to be sought in courses.
   There was, at times, an assumption in discussions that breadth could be maintained at secondary and freshman level and depth would only become a serious issue late in the undergraduate course. This *may* be true for some countries where choice of speciality is delayed through a student's school and university career and a general education is encouraged. It is not true for many European countries and for those less-developed countries which use a European education model.
   It was recognized that it was essential for courses to allow career choice and to permit progress to further education, and it was appreciated that employers would continue to look for specialists, in this instance specialists who have an understanding of global change science.

3. There is much evidence that global change science courses are popular both with students and with staff. The courses are highly motivating, and even

students who have an in-built fear or dislike of quantitative scientific concepts find, because their interest is held, that they are able to understand the principles. Those university departments that advocate single subject courses should take note. Probably they will discover that collaboration to produce multidisciplinary courses won't be at a high cost to themselves. They may even get more students to teach!

4. All courses need new resources for teachers, often in the form of student activities, so that the students can take a positive role in their own learning. Teachers need help with organising teaching strategies to deal with this. Some examples of good practice were given in the papers.

5. Pervading the discussions in every group was the importance of what were termed outreach courses. Particularly important are those in which school teachers are given help with teaching global change science, thus, in turn, helping the next generation to be able to participate more securely in the inevitable debate on global change issues. This will become increasingly important in the next decades, when less developed countries such as China and India, which also have significantly larger populations, will be approaching a standard of living comparable to that enjoyed now by more developed countries.

References

Anastasi C (1994) This book
Boardman J (1994) This book
Camoes MF; Neves LAT (1994) This book
de Groot RS (1994) This book
Moore JC and Lange MA (1994) This book
Schneider S (1994) This book
University of York (1995) Graduate Prospectus, 1995 Entry
White J (1994) This book

NATO ADVANCED RESEARCH WORKSHOP: JUNE 1994

PARTICIPANTS

DR CHRIS ANASTASI
Group Planning, Shell International Petroleum Company, Shell Centre, London SE1 7NA, U.K.

MAX BERAN
TIGER Programme, Institute of Hydrology, Wallingford, Oxfordshire, OX10 8BB, U.K.

DR JOHN BOARDMAN
Environmental Change Unit, University of Oxford, Oxford, OX1 3TB, U.K.

PROFESSOR FILOMENA CAMOES
Department of Chemistry, University of Lisbon, 1700 Lisbon, Portugal.

LYNNE M CARTER
Graduate School of Oceonography, University of Rhode Island, Narragansett, RI 02882, U.S.A.

DR RUDOLF S DE GROOT
Wageningen Center for Environment and Climate Studies,
Wageningen Agricultural University, PO Box 9101, 6700 HB Wageningen, Netherlands.

DR MONTSERRAT FILELLA
Department of Inorganic, Analytical & Applied Chemistry, University of Geneva,
Quai Ernest Ansermet 30, CH-1211 Geneva 4, Switzerland

PROFESSOR GEORGE W FISHER
Department of Earth and Planetary Sciences, The John Hopkins University, Baltimore, MD 21218, U.S.A.

PROFESSOR WILLIAM S FYFE, FRS
Department of Earth Sciences, University of Western Ontario, London, Ontario, N6A 3K7, Canada.

PROFESSOR CATHERINE GAUTIER
Earth - Space Research Group, Computer Systems Lab, Ellison Hall, University of California, Santa Barbara, CA 93106, U.S.A.

KATHIE GERSTER
Oakridge S.S., 1040 Oxford Street, London, Ontario N6H 1V4, Canada

DR HENRIETTA LIDIARD
Environmental Geochemistry Research Unit, Department of Environmental Sciences,
University of Bradford, Bradford, West Yorkshire BD7 1DP, U.K.

DR JEFFREY J MCDONNELL
College of Environmental Science and Forestry, State University of New York, Syracuse, NY 13210, U.S.A.

DR JOHN C MOORE
Arctic Centre, University of Lapland, FIN - 96101, Rofaniemi, Finland.

DR MICHAEL POWELL
Department of Earth Sciences, University of Western Ontario, London, Ontario N6A 3K7, Canada

DR STEPHEN SCHNEIDER
Department of Biological Sciences, Stanford University, Stanford, CA 94305-5020, U.S.A.

DR SERENA SCHWARTZ
Departments of Bio & Atmospherics, Oceanic & Space Sciences (AOSS), University of Michigan, Ann Arbor, MI, U.S.A.

DR PANAYOTIS A SISKOS
Department of Chemistry, University of Athens, Panepistimiopolis 15771, Athens, Greece

DR HENRIK SKOV
National Environmental Research Institute, DK 4000 Roskilde, Denmark

PROFESSOR RICHARD SOMERVILLE
Climate Research Division, Scripps Institution of Oceonography, University of California, San Diego, La Jolla, CA 92093 - 0224, U.S.A.

DR JOSEPH P STOLTMAN
Department of Geography, Western Michigan University, Kalamazoo, MI 49008, U.S.A.

DR DIRK G TROOST
Coastal and Marine Sciences, UNESCO, 1 Rue Miollis, 75732 Paris Cedex 15, France.

PROFESSOR SERGE G TOUCHINSKI
Department of Environmental Management, Faculty of Geography, Moscow State University, 119899 Moscow, Russia.

PROFESSOR DAVID J WADDINGTON
Department of Chemistry, University of York, York, YO1 5DD, U.K.

DR JAMES W C WHITE
Department of Arctic and Alpine Research, University of Colorado, Boulder CO 80309, U.S.A.

The following wrote papers for the meeting but were unable to attend:

DR NANCY HEALY-WILLIAMS
Department of Geological Sciences, University of South Carolina, Columbia, SC 29208, U.S.A.

PROFESSOR WALTER LEAL FILHO
European Research Centre on Environmental Education and Development, Department of Environmental Sciences, University of Bradford, Bradford, BD7 1DP, U.K.

DR DOUGLAS WILLIAMS
Department of Geological Sciences, University of South Carolina, Columbia, SC 29208, U.S.A.

WORKSHOP COMMITTEE

MAX BERAN

PROFESSOR WILLIAM FYFE

PROFESSOR STEPHEN SCHNEIDER

PROFESSOR DAVID WADDINGTON

The ASI Series Books Published as a Result of Activities of the Special Programme on Global Environmental Change

This book contains the proceedings of a NATO Advanced Research Workshop held within the activities of the NATO Special Programme on Global Environmental Change, which started in 1991 under the auspices of the NATO Science Committee.

The volumes published as a result of the activities of the Special Programme are:

Vol. 1: **Global Environmental Change.** Edited by R. W. Corell and P. A. Anderson. 1991.
Vol. 2: **The Last Deglaciation: Absolute and Radiocarbon Chronologies.** Edited by E. Bard and W. S. Broecker. 1992.
Vol. 3: **Start of a Glacial.** Edited by G. J. Kukla and E. Went. 1992.
Vol. 4: **Interactions of C, N, P and S Biogeochemical Cycles and Global Change.** Edited by R. Wollast, F. T. Mackenzie and L. Chou. 1993.
Vol. 5: **Energy and Water Cycles in the Climate System.** Edited by E. Raschke and D. Jacob. 1993.
Vol. 6: **Prediction of Interannual Climate Variations.** Edited by J. Shukla. 1993.
Vol. 7: **The Tropospheric Chemistry of Ozone in the Polar Regions.** Edited by H. Niki and K. H. Becker. 1993.
Vol. 8: **The Role of the Stratosphere in Global Change.** Edited by M.-L. Chanin. 1993.
Vol. 9: **High Spectral Resolution Infrared Remote Sensing for Earth's Weather and Climate Studies.** Edited by A. Chedin, M.T. Chahine and N.A. Scott. 1993.
Vol. 10: **Towards a Model of Ocean Biogeochemical Processes.** Edited by G. T. Evans and M.J.R. Fasham. 1993.
Vol. 11: **Modelling Oceanic Climate Interactions.** Edited by J. Willebrand and D.L.T. Anderson. 1993.
Vol. 12: **Ice in the Climate System.** Edited by W. Richard Peltier. 1993.
Vol. 13: **Atmospheric Methane: Sources, Sinks, and Role in Global Change.** Edited by M. A. K. Khalil. 1993.
Vol. 14: **The Role of Regional Organizations in the Context of Climate Change.** Edited by M. H. Glantz. 1993.
Vol. 15: **The Global Carbon Cycle.** Edited by M. Heimann. 1993.
Vol. 16: **Interacting Stresses on Plants in a Changing Climate.** Edited by M. B. Jackson and C. R. Black. 1993.
Vol. 17: **Carbon Cycling in the Glacial Ocean: Constraints on the Ocean's Role in Global Change.** Edited by R. Zahn, T. F. Pedersen, M. A. Kaminski and L. Labeyrie. 1994.
Vol. 18: **Stratospheric Ozone Depletion/UV-B Radiation in the Biosphere.** Edited by R. H. Biggs and M. E. B. Joyner. 1994.
Vol. 19: **Data Assimilation: Tools for Modelling the Ocean in a Global Change Perspective.** Edited by P. O. Brasseur and J. Nihoul. 1994.

Vol. 20: **Biodiversity, Temperate Ecosystems, and Global Change.**
Edited by T. J. B. Boyle and C. E. B. Boyle. 1994.

Vol. 21: **Low-Temperature Chemistry of the Atmosphere.**
Edited by G. K. Moortgat, A. J. Barnes, G. Le Bras and J. R. Sodeau. 1994.

Vol. 22: **Long-Term Climatic Variations – Data and Modelling.**
Edited by J.-C. Duplessy and M.-T. Spyridakis. 1994.

Vol. 23: **Soil Responses to Climate Change.**
Edited by M. D. A. Rounsevell and P. J. Loveland. 1994.

Vol. 24: **Remote Sensing and Global Climate Change.**
Edited by R. A. Vaughan and A. P. Cracknell. 1994.

Vol. 25: **The Solar Engine and Its Influence on Terrestrial Atmosphere and Climate.**
Edited by E. Nesme-Ribes. 1994.

Vol. 26: **Global Precipitations and Climate Change.**
Edited by M. Desbois and F. Désalmand. 1994.

Vol. 27: **Cenozoic Plants and Climates of the Arctic.**
Edited by M. C. Boulter and H. C. Fisher. 1994.

Vol. 28: **Evaluating and Monitoring the Health of Large-Scale Ecosystems.**
Edited by D. J. Rapport, C. L. Gaudet and P. Calow. 1995.

Vol. 29: **Global Environmental Change Science: Education and Training.**
Edited by D. J. Waddington. 1995.

Vol. 30: **Ice Core Studies of Global Biogeochemical Cycles.**
Edited by R. J. Delmas. 1995.

Springer-Verlag and the Environment

We at Springer-Verlag firmly believe that an international science publisher has a special obligation to the environment, and our corporate policies consistently reflect this conviction.

We also expect our business partners – paper mills, printers, packaging manufacturers, etc. – to commit themselves to using environmentally friendly materials and production processes.

The paper in this book is made from low- or no-chlorine pulp and is acid free, in conformance with international standards for paper permanency.